T0093705

Nature's Wild Ideas

Nature's WILD IDEAS

How the Natural World Is Inspiring Scientific Innovation

KRISTY HAMILTON

GREYSTONE BOOKS
Vancouver/Berkeley/London

22 23 24 25 26 5 4 3 2 1

Greystone Books Ltd.
greystonebooks.com
Cataloguing data available from Library and Archives Canada
ISBN 978-1-77164-819-6 (pbk)
ISBN 978-1-77164-820-2 (epub)

Editing by Linda Pruessen
Copy editing by Jess Shulman
Proofreading by Meg Yamamoto
Indexing by Stephen Ullstrom
Cover design and composite by Belle Wuthrich and Jessica Sullivan
Cover composite illustration credits:
Simple Line, Mikhail Gnatuyk, Crazy nook, jamestar,
Disa_Anna, ujiartdesain, ksanask.art / Shutterstock
Text design by Belle Wuthrich

Printed and bound in Canada on FSC® certified paper at Friesens.
The FSC® label means that materials used for the product have been responsibly sourced.

Greystone Books gratefully acknowledges the Musqueam, Squamish, and Tsleil-Waututh peoples on whose land our Vancouver head office is located.

Greystone Books thanks the Canada Council for the Arts, the British Columbia Arts Council, the Province of British Columbia through the Book Publishing Tax Credit, and the Government of Canada for supporting our publishing activities.

Canada

To my family and friends.
Although many of us may not leave riches to our families, we can bequeath them a legacy more precious than gold, more delicate than glass, and more monumental than fame: a world preserved in the amber of our protection.

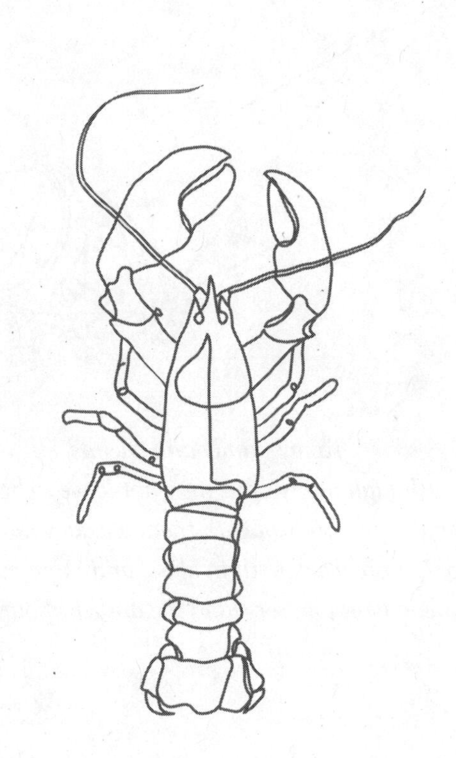

Contents

Introduction

WILDLY INSPIRING

IN 1874, INVENTOR Alexander Graham Bell was a twenty-seven-year-old with a dark, bushy beard working on a world-changing invention, and he was doing so by peering at the swirls of a cadaver's ear, an exquisite fleshy instrument that has taken millions of years to develop. Bell was astonished by how the ear's thin membrane could move the weight of the middle ear bones: "It occurred to me that if a membrane as thin as tissue paper could control the vibration of bones that were, compared to it, of immense size and weight, why should not a larger and thicker membrane be able to vibrate a piece of iron in front of an electromagnet?"[1] Beside a sketch in his notebook, he scribbled down, "Make transmitting instrument after the model of the human ear. Make armature the shape of the ossicles [small bones in the human ear]. Follow out the analogy of nature."[2]

Bell had good reason to explore this line of inquiry: both his mother and his wife were deaf, and he taught students with hearing impediments. When he built his ear phonautograph, he used the actual bones of a human ear mounted on a wood frame. Voices caused the bones to vibrate and visually represent the sound waves as etchings on a smoked-glass plate. The invention was meant to be a tool for his deaf students, but it leapfrogged his imagination to insight that eventually led to his telephone patent in 1876. Sitting in his laboratory, he famously spoke to his assistant on the phone in another room and said, "Mr. Watson, come here. I want to see you."[3]

History is rich with such examples: from penicillin derived from fungi to cancer drugs developed from coral to painkillers inspired by venomous frogs and cone snails. Biomimicry (from the Greek *bios*, meaning "life," and *mimesis*, meaning "to imitate") is the study of nature and takes inspiration from its magnum opus of ideas. The first person to coin the term *biomimicry* (also called *bioinspiration* and *biomimetics*) was biophysicist Otto Schmitt, in the 1950s. His idea was popularized in 1997 by biologist and author Janine Benyus, who believes scientists should be at the design table too. Biomimicry is still a relatively nascent discipline, and it's important to discern fact from fiction, especially when biomimicry is harnessed as a marketing tool and not as an investment in scientists spending years to gain new insights. Biomimicry is also not a cure for all that ails us—it is a guiding light, a source of inspiration, a place of awe where we can marvel at ideas we never conceived of ourselves. In writing this book, my hope is to peer behind the curtains of biomimicry and explore the detective work that has occupied thousands of scientists around the world—men and women who are churning the metaphorical crank of this field ever faster. All around us, creatures millions of years in the making are harnessing energy and materials; they don't produce pollution like us, but they do evolve ingenious solutions to hack their way to survival.

Nature's Wild Ideas is about the animals and plants that have inspired everything from telescopes to view cataclysmic explosions in the universe to medication for hard-to-treat diabetic patients to a prize-winning discovery that, according to the Nobel Foundation, has become "one of the most important tools used in contemporary bioscience."[4] We can even find "human inventions" already in use in the animal kingdom: electric eels generate electricity powerful enough to stun a person; squids use jet propulsion; tree crickets turn leaves into mini-megaphones to amplify their calls; and beavers build dams to flood lakes for safe housing and movement. The swim bladder of fish helps them control their buoyancy, similar to the ballast tanks of submarines. Even the agricultural revolution wasn't that, well, revolutionary. Several ant species have long known how to cultivate fungus gardens and herd aphids for their "milk," gently stroking them with their antennas to release droplets of honeydew, a sweet fluid they excrete after feasting on sap or leaves. A humpback's throat is, in essence, large-scale origami, but with folds of skin rather than factory-made paper, expanding thanks to as many as thirty-six grooves that stretch to capture prey and collapse into a compact form when done.

This book is an in-between art: part discovery, part science, part natural world, part philosophical questioning. What is the natural world to us? How important is it that we preserve the diversity of life on Earth? What is our role in the tangled web of creation and innovation? In many ways, nature is more elusive to mimic than we ever imagined. Often we find ourselves trying to extrapolate from the wild and invent something never before seen in the natural world. It takes an interdisciplinary pack of biologists, engineers, chemists, physicists, materials scientists, mathematicians, and more to come together and see the potential. Like explorers poring over the map of an obscure place and wondering what lies in the empty spaces, scientists dig deep into unknown, undrawn places and try to pioneer new insights to add to our collective knowledge. Institutions across the world have added departments solely dedicated to the endeavor of biomimetic science: Massachusetts Institute of Technology, the Wyss Institute at Harvard, Georgia Tech's Center for Biologically Inspired Design, Imperial College London's Centre for Bio-Inspired Technology, and the Biomimicry Center at Arizona State University are but a few.

As a last point of note, I'd like to clarify from the outset that evolution does not have foresight or some divine plan in mind; it is a series of adjustments to adapt to local changes and the environment. As such, biomimicry is not meant to be taken as an end-all, be-all; it serves as inspiration within a set of limitations. Evolution does not have an inventive mind like engineers, and animals have biological constraints such as the need to eat, reproduce, and defecate—necessities our products and machines can go without. However, biological designs can provide fresh solutions to old paradigms. For example, how do blue mussels create a glue that withstands the wet battering of waves? Or why doesn't blood pool in a giraffe's slender legs, given the creature's lofty height?

The inventions mentioned in this book are by no means perfect in the whole sense of the word, but they do inspire the imagination toward something greater, deeper, and more symbiotic with the world around us. Unfortunately, humanity's connection to nature is only diminishing with time. In just one generation, time spent playing outdoors in the United Kingdom has halved, according to National Trust, with many kids spending the same time outdoors as stipulated in the UN's guidelines for prisoners: "at least one hour of suitable exercise in the open air daily if the weather permits."[5] To put it simply: civilization has lost touch with the wild, and yet we crave the elemental, the tangible, something that feels like more than just our screens and offices. And why shouldn't we? Nature is deeply rooted in our humanity. If we are to continue learning from the creatures around us, we must protect their untamed lands. Just as we have the power to design our life with the arbitrary givens handed down to us, we as a species have the power to design a better future.

I hope you will join me as we traverse frozen waterfalls, trek through cloudy forests, and scour intertidal zones to discover *Nature's Wild Ideas*.

1

A COLD CASE

Tardigrades Inspire the Preservation of Medicine

"Nothing burns like the cold."

GEORGE R. R. MARTIN, *A Game of Thrones*

FIFTEEN MILES SOUTHEAST of Bozeman, Montana, is a valley nestled into the northern section of the jagged, snow-crusted Gallatin Range. Come winter, subzero temperatures freeze the valley's hundreds of waterfalls into spears of ice, the flow of water that once roared from the cliffs now chilled into stillness. If you squint, pops of scarlet or canary yellow speckle the icefalls. I'm one of the novice ice climbers, wearing an electric-blue jacket as I swing a steel ice ax over my head. *Thunk.* The hit sends vibrations down my forearm. Flakes of ice crash from above and burn cold on my face. I kick my boots, rigged with crampons, into the frozen waterfall as my breath vaporizes like chimney smoke.

Thunk. More ice chips tumble to a carpet of snow below, a world of microorganisms within. For most of human history, we didn't know it was feasible for life to shrink out of sight and still possess a brain. Equally implausible was that the tiniest of tiny creatures could survive in the ice. Then, in the seventeenth century, a secretive man with brown barrel curls and a pencil-thin mustache revealed just how blind we are to much of the planet. Anton van Leeuwenhoek didn't intend to change our view of the world or become immortalized as the father of microscopy; no, he was simply trying to assess the quality of the threads at his shop. Using his talent for making lenses, he heated thin filaments of glass into small spheres to construct a microscope. Out of curiosity, van Leeuwenhoek aimed his microscope at the scummy water he scooped out of a pond and the white stuff ("as thick as wetted flower") scraped off his own teeth.[6] What he witnessed was a sheer thrill. Who truly gets to say they have discovered a new world, like the hero in some childhood fantasy? Van Leeuwenhoek saw "small living animals, which moved themselves very extravagantly." He was also the first to see red blood cells, sperm (taken from his own marital bed), and the vast unseen life of Earth's littlest residents, which he called "animalcules." And yet, because he refused to share his lens-making methods, no one but van Leeuwenhoek could see this microcosmos he kept jabbering about. He wrote to Robert Hooke, an English scientist and architect, "[I] oft-times hear it said that I do but tell fairy tales about the little animals."[7]

We now know van Leeuwenhoek's writings were not fairy tales; in just a teaspoon of soil there are a billion bacteria (around the same number of humans living in the Americas), thousands of protozoa, and scores of nematodes and fungal filaments. Our own bodies contain billions of single-celled organisms, many of them our allies when it comes to digesting food and ridding us of illness. We even excrete our own weight in fecal bacteria every year. Three hundred years after van Leeuwenhoek's discovery, we have vanquished microorganisms with our medicine, transplanted fecal bacteria from one person to another to treat *Clostridioides difficile* (a bacterium that can cause life-threatening diarrhea), and even engineered bacteria to do our bidding (to, say, kill parasites). And yet, despite such advances, these organisms still have the power to baffle us. Consider, for example, a creature that could very well be living in a state of suspended animation inside the ice I've dislodged on my climb—a creature unlike any other on the planet.

If you want to meet the ultimate survival artist, look no further than the tardigrade (pronounced TAR-dee-grade)—a.k.a. water bear, a.k.a. moss piglet. The multiple monikers are likely due to the creature's debatable looks: Does it resemble a bear or a pig? Is it adorable or hideous? To me, it's all of these things. Imagine a microscopic eight-legged gummy bear with wickedly curved claws, but where there should be a face with a mouth and eyes, there is a piggish snout instead. That's a tardigrade. Freeze these creatures in a cryogenic tomb, dry them out for a hundred years, or zap them with high doses of radiation, and guess what? They will survive. Tardigrades have graced Earth for about 600 million years, which is more than 350 million years before dinosaurs and flowering plants, and they have survived all five mass extinctions.

And here I am—a member of a species that's spent a piddling 200,000 years on Earth—stuck to the side of a waterfall in the middle of winter with "pocket warmers" stuffed into my gloves and socks. I am, it is safe to say, no tardigrade. We humans are delicate, temperate-climate-loving creatures. Our survival has less to do with our physical abilities and more to do with our talent to devise tools to overcome our mortal limitations. We've done pretty well too, mimicking nature's cold conditions by developing artificial refrigeration in the mid-1700s. This is relatively late in our history compared with heat because we're better at creating warmth, like fire, than cold. And herein lies the conundrum that had us stumped for so many years: To create heat, you produce it. But cold is the *absence* of heat, and removing heat is much harder than creating it. We have now gotten so good at manufacturing the perfect temperature for whatever we need and wherever we go—be it a wine cooler in South Africa or ice cubes in Palm Springs, California—that when this cold stability breaks down, we're left utterly helpless. The ability to keep temperatures stable—a process called the "cold chain"—is a life-saving necessity for most of the world. We depend on this cold chain for everything from refrigeration for our food to scientific experiments and medicines. Insulin for diabetes, Humira for arthritis, and Epogen for chronic kidney disease are just a few of the biological drugs that need to be kept in a stable, cold environment. This is also true for all vaccines.

Vaccines are sensitive to temperature extremes. If you live off the beaten path, it can be tough to get them to you before they expire. In parts of Kenya, for example, vaccines are transported to villages inaccessible by car using the local equivalent of a mobile clinic: camels carrying solar-powered mini-refrigerators on their humps. At all times, freezers and cold packs must maintain a constant temperature. Despite this effort, a third of all refrigerated vaccines and pharmaceuticals shipped to developing countries are cracked or degraded by the time they reach their final destination. Children all over the world die of viral infections that can be cured.

Viruses were discovered only 120 years ago, around the same time as human flight and plastic. This is astounding when you consider that more than a quadrillion quadrillion viruses exist on our planet, a greater number than the stars in our Milky Way. In just a drop of seawater, there are thousands, even millions, of viruses. If all of Earth's viruses were lined up end to end, they would tower higher than the sun, past Pluto, and beyond the Andromeda galaxy for 100 million light-years.[8] Some 380 trillion viruses live inside each one of us, outnumbering bacteria ten to one. Thankfully, only a select group are dangerous to humans. Upwards of two hundred viruses can break into human cells and cause disease. These viruses are some of the strangest entities on Earth—they are not alive and they are not dead either; they carry DNA like the living but they cannot reproduce like we can. Instead, viruses must hijack our cells and inject their genetic material inside, commandeering our cells' ability to replicate. These invisible saboteurs can give rise to conditions like encephalitis (swelling of the brain), hemorrhagic fever, the common cold, hepatitis, or skin lesions, or symptoms like those experienced with COVID-19, which hits humans where we are weakest: our desire for community and connection. Our spoken words coming from the very same portal through which the deadly infection makes its insidious entry.

Medicine is a key tool in the fight against these unseen threats, but an estimated 19 million children younger than one year of age do not receive their basic vaccinations every year. A key barrier in getting vaccines to these children is the cold chain. To understand why, we need to delve into the vaccine vial, where a virus usually lies in either a weakened or an inactivated state. Most vaccines require a stable temperature in the range of 35.6 to 46.4 degrees Fahrenheit (2–8°C). Dip below this perfect temperature range, or rise above it, and the virus membranes and DNA start to break down. This makes transporting the treatments to remote regions of the world logistically challenging; to do so requires meticulous temperature control and strategic expertise.

This situation is nothing short of tragic in a civilization that has come so far in medicine. How can something as seemingly simple as temperature control be standing in the way of positive health outcomes for millions of people? Isn't there a way to preserve life-saving medications so that they reach their intended recipients in time to make a difference? The solution may lie in the most unlikely of places.

Small Wiggly Things

I'M AT A banged-up desk, twisting the knob of a microscope to focus the lens on a pond sample in a petri dish. As I squint into the instrument's eyepiece, a blurry blob focuses into a tardigrade as big as a beetle. As many times as I've heard of these creatures, there is nothing like seeing one for myself. I'm suddenly left with the feeling that we might be living inside a Matryoshka doll, where there is a world within a world within a world—whole ecosystems of creatures oblivious to the vast echelons of sizes that exist.

In this moment, I'm a voyeur of the tardigrade's micro-realm, watching as green juices from a meal of moss wriggle through the creature's gut. As I observe I chat with Mark Blaxter, who taught at the University of Edinburgh for twenty-five years while researching "small wiggly things" like roundworms and tardigrades. Blaxter is now in charge of the Tree of Life Programme at Britain's Wellcome Sanger Institute. His grand goal is to sequence all species on Earth, but first the program is starting with eukaryotic species—protists, fungi, plants, and animals—in the British Isles. It's a difficult but not impossible mission. And Blaxter likes extremes. His enthusiasm for small wiggly things is even evident in what some might call the most sacred realm of one's home: the fridge. Stashed inside his is a fourteen-year experiment in a Ziploc bag.

"They're very cute tardigrades, these ones," he says of his own bagged beasties, his gray hair tied in a low ponytail at the base of his neck. "They're called *Ramazzottius*. They're a nice, bright pink color. We're getting fewer every year, so they're not going to last forever."[9] (Though they've certainly outlasted the pasta.)

If you've managed to keep a creature in your fridge for longer than most Americans remain married, you probably know a thing or two about it. Tardigrades are found on every continent and at nearly all elevations: in deep-sea trenches, burbling hot springs, forest canopies, and desert dunes. They are so ubiquitous, in fact, that you've probably swallowed one in a gulp of water. Every so often, they even catch a flight to new lands, courtesy of birds. Tardigrades have no need for jackets or pocket warmers to stay alive in the frigid depths of space or the snow flurries of mountain peaks. They can survive for minutes at a blistering 304°F (151°C) or for a few days at a bone-chilling –328°F (–200°C). They were around when crocodiles lounged under palm trees in the Arctic 56 million years ago, and when Antarctica was a swampy rain forest 90 million years ago.

Tardigrades, however, are not extremophiles; they don't seek out extreme conditions in which to live. It's easy to give them capes and call them superheroes, but they prefer a cozy environment with food and fresh air. Unlike humans, they are good at withstanding environmental extremes when their homes disappear. And yet, they're rather simple creatures. Most tiny invertebrates zip around a dish like pint-sized rockets, but pudgy tardigrades mosey along on stubby legs (hence their name *Tardigrada*, from the Latin word for "slow walker"). They don't have a circulatory or respiratory system; their body's open cavity allows nutrition and gas to touch every cell and sustain life.

The two most famous tardigrades in existence were frozen in a thirty-year cryogenic pause until they were thawed and their resurrection captured on tape. The samples were collected in 1983 in the Yukidori Valley of Dronning Maud Land, Antarctica, where a team from the National Institute of Polar Research gathered frozen moss samples containing tardigrades, wrapped them in paper, placed them in plastic bags, and—with the utmost of scientific patience—stored them at −4°F (−20°C) for the next 30.5 years.

The tardigrades were dubbed "sleeping beauties" by news outlets all over the world. During their deep freeze, humans invented the World Wide Web and launched the Hubble Space Telescope; Nelson Mandela was freed from prison and won the Nobel Peace Prize; tobacco companies admitted to the harm of cigarette smoke; and the United States entered a war with Iraq, twice. Then, in 2014, in what must be their equivalent of an earthquake *and* a tidal wave, the samples were picked up, thawed, and soaked for twenty-four hours. The tardigrades were suctioned up with a pipette and inspected under a microscope by a massive, curious eye. Only two tardigrades and one egg survived—but, most importantly, they did. A twitch in Sleeping Beauty 1's leg the first day, another twitch the next. It took the tardigrades a couple of weeks to fully revive (not the "kiss and all's well" of their namesake princess), but as they did, humanity became a witness to an implausible feat of endurance. By the third week, three eggs were visible inside Sleeping Beauty 1 and another generation was born. All was well.

How was this thirty-year deep freeze possible? Scientists believe that in order to freeze safely, tardigrades dry up via a feat called *cryobiosis* (from the ancient Greek *krúos*, meaning "icy cold," and *bíosis*, meaning "way of life"). This drying up is vital to their self-preservation. If you've ever left a water bottle in a freezer and later gone to retrieve it, you will have noticed that the bottle swelled in girth. When water freezes, it takes up more space; in organisms, the sharp tips of ice crystals are like switchblades that can rupture cells, puncturing membranes and DNA. Tardigrades don't need to worry about this because when they curl into a ball, they shrivel up like raisins and lose up to 97 percent of their water. In this dried state, called a tun, they can last for decades, their cells intact. It's almost as if they wilt into a powdered version of themselves.

For humans, such a lack of water would spell disaster. Everything we need to keep wet (like our gut, lungs, and brain) is inside our bodies, which are 60 percent water. We've evolved special mechanisms to protect our fragile organs and cells from getting too dry. Our skin prevents all our water from immediately evaporating, while still allowing sweat to seep from our pores to cool us down (like an internal sprinkler system). The big fleshy appendage on our face also plays a part. Our nose processes the air we breathe and prepares it for the lungs, which do not tolerate dry air well; it's like a portable humidifier, warming and moisturizing the air before it goes to the lungs. You can imagine what would happen to us if we were to dry up. Our lips would crack and bleed, our throat would throb from the dry air, the mucus that lines our airways and lungs would thicken and get stickier, fatigue would set in, and confusion would emerge. All of our enzymes and DNA are effectively based in water, so they would begin to lose their three-dimensional shape before the body's catastrophic collapse. Humans last about four days without water. Camels in the Sahara Desert live for a week. Tardigrades can survive for thirty years, similar to their survival in ice.

If such resilience is possible, why is the tardigrade one of the few creatures on Earth to possess it? Researchers believe the answer lies in the tardigrade's evolutionary history. The story begins like that of all other major animal groups on the planet: tardigrades came from the sea and only after millions of years did they begin the daring journey of life on land. But it was risky. Unlike other animals, tardigrades don't have waterproof skin to hold water in their bodies. Instead, they evolved the ability to enter a tun state, a dormant period where their organs and cells are safely packed until enough water returns to bring them back to full life. This ability also made them resilient to other harsh conditions—like the cold, radiation, and the vacuum of outer space—because inadvertently the tun state works well for these stresses too.

"Anything that lives in the desert or a habitat that is sometimes wet, like an ephemeral pond, and then dries up has a special problem because they have to really hang on to water," says Blaxter. "So those organisms have evolved 'superpowers,' or whatever you want to call it, that allow them to live without water. Tardigrades are famous for this because they are ubiquitous and live in environments that rapidly dry up and get wet again."

Even if we humans could endure losing most of the water in our bodies (which we definitely cannot), we would still perish from damage to our DNA and proteins. If we rehydrated like tardigrades, nothing inside our bodies would function. A tun state is not simply a matter of packing all the organs and cells closer together. Tardigrades also have to protect their insides from clotting, tangling, or rubbing together in damaging ways. When tardigrades return to life, everything in their body needs to be in working order, including their one thousand cells. This knowledge guided researchers to their next questions: What solutions are tardigrades using to perform this seemingly miraculous feat of self-preservation? If tardigrades don't want to experience a catastrophic collapse of their bodies, they've got to replace the water with something else—something that won't evaporate. But what?

Life Is Like a Peach

PROTEINS ARE LIKE peaches, says Judy Müller-Cohn. If you freeze peaches, defrost them, and freeze them again, they don't hold their shape. The same is true for proteins in laboratory samples. An accidental freeze-thaw cycle—if, say, the cold chain breaks down—can degrade them. Proteins are molecules linked in a chain and twisted into a variety of shapes. Each type of protein has a preferred way to fold, a bit like an origami project. An insulin protein, for example, clumps into a ball shape, whereas a collagen protein prefers a tight triple helix. Most proteins need to stay in their preferred shape to work. A watery environment helps proteins to do this; in fact, if water is removed, most proteins unfold and break apart.

Müller-Cohn faced the problem of the cold chain firsthand when she was working as an entomologist at a retail seed company that develops corn hybrids. "One night, the freezer went down and we lost a lot of samples. I was called at two in the morning because the samples were going bad in this defrosting freezer," she recalled. "Biological samples are kind of our currency in biotech. The nucleic acids, the proteins, cell lines, they're what we do to discover new technologies, new medicines, and new diagnostic applications."[10]

Müller-Cohn was a tree planter in the Northwest before changing careers to work on ways to cut down on chemical insecticides in the environment. With her husband, Rolf Müller, she founded biotechnology company Biomatrica. The duo are molecular biologists whose paths merged at Oregon State University. They continued their work in Paris, earning doctorates at the Pasteur Institute and then landing careers in bioscientific research. Her freezer breakdown was a stark reminder that although DNA is a rather stable molecule, RNA is not. (This is due to differences in their chemical composition, such as RNA being single-stranded and DNA being double-stranded.)

Is there a way to store biological samples at room temperature? Müller-Cohn wondered. If so, it would cut down on the carbon footprint of scientific research and lead to better accessibility—a win-win scenario. But it's a boulder of a problem to overcome. I've witnessed firsthand how easily the cold chain can break down. In 2019, I was at the Woods Hole Oceanographic Institution in Falmouth, Massachusetts, to interview scientists when a "bomb cyclone" hit the region. Thunder roared like an old dry laugh and gusts as powerful as eighty miles per hour hit nearby towns, felling power lines and blowing out all the lights. Trees were uprooted, and lanterns and woolen blankets were handed out to those of us stuck in the dark. We all flocked to a café with a backup generator, called Pie in the Sky, to eat and charge our electronics. All interviews were canceled until the winds died down, power was returned, and the roads cleared. This didn't take hours. It was days. Whether any scientific samples were lost due to the power failure I don't know, but the episode does raise the question of our precarious dependence on the cold chain.

Müller-Cohn worried about scenarios like this. As a scientist herself, she knows all too well how much time an experiment can take; they are often years in the making and require hundreds of thousands of dollars in funds. This got her thinking: Are there creatures that survive these kinds of high-stress situations? After some research, she landed on the tardigrades and their ability to slow perishing in life's harshest conditions. Something inside their bodies requires time to make protectants, "so we looked at the chemistry of it and we tried to copy it," she says. "It was true biomimicry."

Some of the research Müller-Cohn likely read was published by another scientific pair, John and Lois Crowe, who pioneered the study of tardigrade survival. John was a professor of molecular and cellular biology at UC Davis before he retired, and Lois was a biophysicist at the same institution. Together they zeroed in on a powerful sugar in tardigrades called trehalose. A similar sugar is found in brine shrimp (also known as sea monkeys), and it's the key to how they survive drying out. Under normal conditions, trehalose is floppy and shapeless, but when the water wanes, it hardens like glass and protects the cellular components. The trehalose holds everything in an internal cast, preventing the cells from shattering or fusing into mush. In effect, the tardigrades make their own internal shield. When tardigrades in a tun state are soaked in a drop of water, trehalose melts like a sugar cube in a warm cup of tea and becomes shapeless again. This frees the cells from their matrix and allows them to resume normal function. Another way to think about trehalose is that it forms a super-viscous environment that slows everything down, almost like honey. With the passing of time, damage like proteins breaking may still occur, but only very slowly, over decades. So as long as tardigrades don't remain in their tun state for too long, they will be okay.

The Crowes' preservation discoveries have ricocheted into other fields and influenced the lives of immune-suppressed patients with life-threatening fungal infections. Prior to their work, there was potential commercial value in liposomes, biological bubble sacs that can encapsulate and deliver drugs to a target site in the body. However, there were issues with the liposomes fracturing upon rehydration and spilling their contents. The Crowes invented a method that made sure there was a good concentration of trehalose inside and outside the liposomes before freeze-drying. As their patent in 1989 notes, now liposomes "can retain as much as 100% of their original encapsulated contents upon rehydration."[11] The Crowes' method was licensed by the University of California to Vestar Research and acquired by

Gilead Sciences for an injectable therapy called AmBisome, which brought in an estimated $1.3 billion in sales from 1986 to 2007.

Inspired by the survival artistry of tardigrades, Biomatrica also developed ways to store tissue, DNA, and cell samples. The technology saves laboratories money and protects research from power outages. Instead of putting materials in a freezer, they can dry them down. In 2018, Biomatrica signed a deal with a diagnostics company to help preserve and transport RNA in blood samples at high ambient temperatures for up to three months. The innovation expands the reach of HIV testing to millions in remote locations and eliminates the unpredictability of cold transportation. "Now HIV blood samples for HIV detection are done on our stabilization matrix," says Rolf Müller.[12]

Trehalose is also used in a variety of commercially available medications, including Herceptin (breast, stomach, and esophageal cancer), Avastin (colorectal, lung, glioblastoma, kidney, cervical, and ovarian cancer), Lucentis (loss of vision from wet macular degeneration), and Advate (hemophilia A, a hereditary bleeding disorder caused by a lack of blood-clotting factor VIII). Trehalose is even found in food and cosmetic products, and there's growing interest in its use for quick-dissolving tablets.

Trehalose is indeed a powerful sugar, but it isn't quite the end of the tardigrade's story.

Planet Medicine

JUST WHEN YOU think tardigrades no longer have the capacity to surprise us, an up-and-coming scientist proves everyone wrong. Thomas Boothby is an assistant molecular biology professor at University of Wyoming, and he's recently been popping up around the world, from TEDx stages to the pages of the *New York Times*. As a young researcher rising to tardigrade fame, Boothby's mission is to study the fundamental mechanisms of extreme stress tolerance (an objective anyone who has spent years getting a PhD can sympathize with). In 2017, Boothby found that although tardigrades do make trehalose, they also make a lot of certain three-protein groups collectively called "tardigrade-specific intrinsically disordered proteins," or TDPs for short. These proteins are not found in any organism except tardigrades, and they hold fascinating potential. Yeast and bacteria that were genetically engineered to have TDPs became one hundred times more resistant to extremely dry conditions than those without the tardigrade upgrade.

Boothby's research suggests TDPs work ten times more efficiently than trehalose and possess greater stability at higher temperatures. He's investigating how to use these proteins to protect biological material, such as medicines and vaccines. The discovery has Boothby energized, and his vision of the future is perhaps grander than most. Simply put, he wants to eliminate our dependence on the cold chain. Boothby has noted that, in developing countries, around 90 percent of a vaccination program's costs are due to the need to keep vaccines cold.[13] But since tardigrades have evolved TDPs to protect tardigrade cells, he has to alter them to work with our medicines. The need for vaccines to be kept cold is one hurdle in the fight against COVID-19, for example; Boothby hopes his work will help preserve them at room temperature. "We have patents out on these things and have some partnerships," he told the *Guardian* in 2021. "If all goes well, hopefully we will see this technology out soon."[14]

The discovery comes at a time when we are trying to reduce our dependence on the cold chain and phase down hydrofluorocarbons (HFCs) in electric refrigerators (HFCs are up to 10,000 times more potent than CO_2). While HFCs are safer for the ozone than previous generations of coolants like R-22, their emissions could rise to 7 to 19 percent of global greenhouse gas emissions by 2050 if society continues as usual, according to the United Nations Environment Programme.[15] In 2009, researchers at Stanford University conducted a pilot project that found if they swapped 2,000 laboratory freezers with room-temperature storage technology, the university's carbon footprint could be reduced by 18,000 metric tons and save $16 million in operating costs over a ten-year period. The feasibility of that shift is yet to be seen, but one point remains: tardigrades react to changing extremes with an incredible command of their cellular nature, but we are not so lucky—and neither are many of the creatures with whom we share this planet. We tend to think of climate change as if it were a distant thunder, but it is in fact precariously near.

With more than a thousand species of tardigrades, there's a lot to explore. The phylum Tardigrada are a crayon box of talents: some tardigrade species survive in cold temperatures (through cryobiosis); others survive in dry conditions (anhydrobiosis), through a lack of oxygen (anoxybiosis), changes in water salinity (osmobiosis), or in the presence of high levels of toxins (chemobiosis).

"Tardigrades have evolved a number of different molecules that stop enzymes from collapsing," says Blaxter. "Some of them look after proteins and stop them from falling apart. Some of them look after lipid membranes because around all of our cells is the cell membrane, which is basically lipids. And some of them, excitingly, look after DNA."

If trehalose and TDPs were not enough, tardigrades also have a protein called damage suppressor (Dsup) that binds to DNA in the cell and protects it from damaging chemicals. The pinkish *R. varieornatus* can withstand a hefty dose of radiation that would kill most other animals on Earth. "Dsup, which actually goes into the nucleus and protects the DNA, wraps around the DNA and stops it from falling apart and being damaged," says Blaxter.

In humans, high doses of X-rays (higher than that used in medical imaging) can cause genetic mutations that lead to illness or death. In one study, human cells that were engineered to produce Dsup saw 40 percent less DNA damage after exposure to X-rays compared with control cells. The finding could be used to help cells live longer in extreme conditions and expand the range of cell applications in biotechnology.

Our knowledge of tardigrades is only the tip of the iceberg. We've rocketed them into outer space, and more than half survived air deprivation. Their near indestructibility has led some to theorize that they could seed life on other planets. Known as panspermia, the theory suggests that hardy life forms like bacteria or tardigrades exist throughout the universe and can be disturbed by meteorites flung up after an asteroid strike, perhaps ending up on another world as stowaways. Scientists tested questions like these by shooting tardigrades out of a high-speed gun and recording whether life or mush remained. They found tardigrades can survive impacts of up to about 1,800 miles (3,000 kilometers) per hour—an incredible act, but likely not enough to endure a planetary impact crash.

Scientists have dismissed panspermia as a fringe theory. It's possible but extremely unlikely. It's more plausible that planets are like biogeographical islands, similar to continents thousands of years before *Homo sapiens* knew the others existed; tardigrades are trapped here, unable to escape the boundaries of Earth. Scientific funding has since shifted away from the panspermia theory to questions of human survival. How do we create a spacecraft that speeds through the cosmos without ill consequences to astronaut health? A study will soon commence aboard the International Space Station to observe how tardigrades adapt to the stresses of spaceflight. NASA hopes the results will offer inspiration for future treatments for astronauts or protections for long-term space missions. Lastly but certainly not exhaustively, Harvard Medical School is investigating whether tardigrades can inspire human therapies that push pause on tissue damage and cell death from traumatic injuries, heart attacks, and other conditions. Researchers from the University of Louisville are similarly working on a trehalose method to convert red blood cells into a powder form for soldiers and trauma patients.

As we learn more of these biological principles, our understanding of nature—and ourselves—continues to evolve. Inspiration from the microcosmos is vast and promising, and tardigrades once again prove that nature is often more outrageous than our categorical minds like to think, especially when a microscopic creature with a squashed snout is endowed with powers we typically reserve for our mythical figures.

2

FISHING FOR STARS

Lobsters Inspire Telescopes to View Cataclysms in the Cosmos

"The eye owes its existence to the light. Out of indifferent animal organs, the light produces an organ to correspond to itself."
J. W. VON GOETHE, German poet

IN 2012, NATIONAL Public Radio called him the Mirror Man—which is fair enough, given that he's spent the majority of his life tinkering with the things. He makes them stronger, bigger, lighter. He rotates them, heats them, makes stuff to clean them. You name it, he does it. It's safe to say that James Roger Prior Angel (who prefers going by Roger Angel) knows a lot about mirrors. But these aren't the kind of mirrors we use to peer at our faces over the sink each morning. No, Angel creates mirrors for telescopes that peer deep into the universe.

It's perhaps fitting that a man named Angel is helping humanity to gaze at the heavens. Angel is an astronomer who talks in a slow, genteel manner and wears thin metal spectacles. His scientific ideas are eccentric in a way that's both expected of him and respected. As Brian Schmidt, a Nobel Prize winner in physics who discovered that the expansion of the universe is accelerating, tells NPR, "Roger's always been someone who has lots of wild ideas; some of them are superb, some of them probably not so good. But he's incredibly creative. That creativity coupled to sort of a genius is quite rare."[16]

It was as a young boy that Angel fell head over heels with observing faraway unknowns and architecting his own creations. He was brought up in a house at the edge of postwar London, England, and if he peered out the back window he could see rows and rows of houses, all the way to the center of the city. But if he ran to the front window, he saw Epping Forest, a place he describes as a free, open area. Nearby, there was a narrow shipping channel, and it was here he built his first telescope. He bought a lens and put it together in a tube so he could read the names of the ships going by. Years later, he earned a doctorate in physics from Oxford and became an associate professor at Columbia University. He went on to receive a full professorship at the University of Arizona and was a recipient of the MacArthur "Genius Grant," becoming an expert on white dwarfs—the smoldering remains of burned-out stars that have exhausted their nuclear fuel and are fated to a slow cooling over the next billion years.

But all things considered, Angel has received little recognition outside scientific circles, and he continues to work underground at the Steward Observatory Mirror Lab in Tucson, which he founded beneath the University of Arizona's 57,000-seat football stadium. To appreciate the importance of Angel's work, we first need to understand a few things about telescopes. The word comes from the Greek *teleskopos*, meaning "far-seeing," and the Oxford Dictionary defines it as an optical instrument designed to make distant objects appear nearer and larger. The definition is correct, if somewhat lacking in nuance. Telescopes are tremendous achievements of knowledge and expertise, gathering not just one wavelength of light but many. Depending on the design, a telescope can collect anything from radio waves to visible light waves to gamma rays to ultraviolet waves. Without telescopes to aid our eyesight, we see but a sliver of the universe, a keyhole glance. The human eye is privy to less than half a percent of the electromagnetic spectrum—a term for all the light that exists in the universe. We are, by all accounts, blind creatures trying to use our staggering creativity to invent ways to help us see more clearly.

We have spent centuries refining our inventions to spy ever farther, both in the most literal sense of distance and in the breadth of the light spectrum. In ancient Rome, Emperor Nero is said to have squinted through two polished emeralds at gladiators in the distance. In 1608, Hans Lippershey applied for a patent for a spyglass that aided "seeing far things and places as if nearby."[17] It was a simple design: a convex and a concave lens in a tube. The spyglass magnified objects three or four times what the unaided eye could see. Yet the government found the invention too easy to copy and vetoed his patent, saying the device had likely already been discovered. Instead, they employed Lippershey to churn out more binocular versions.

A year later, news of Lippershey's invention reached Galileo Galilei in Venice, and he, an expert toolmaker in his own right, immediately went about creating a stronger, better telescope. Galileo was a child of the astronomical revolution, partaking in a tectonic paradigm shift in our view of the heavens. With his telescope he could observe objects in the universe sixty-four times fainter than with the unaided human eye. Never-before-seen moonlight tunneled through the telescope and into Galileo's mind. Four moons circling around Jupiter—an unprecedented discovery that our moon is not the sole satellite in the universe. The dawn of the telescope era transformed the myth of journeying to the heavens into a possibility. In 1610, astronomer Johannes Kepler wrote in a letter to Galileo that "given ships or sails adapted to the breezes of heaven, there will be those who will not shrink from even that vast expanse. Therefore, for the sake of those who, as it were, will presently be on hand to attempt this voyage, let us establish the astronomy, Galileo, you of Jupiter, and me of the moon."[18]

The sole purpose of a telescope is to collect and focus light. As Dr. Geoff Cottrell writes in his book *Telescopes: A Very Short Introduction*, "the most important feature of a telescope is that it should be a good light collector. Like a bucket put out in the rain to collect raindrops, the larger the aperture of the telescope, the more photons it can gather. But a telescope also has to concentrate, or focus, the light to create an image."[19] The first telescopes used glass lenses to focus the universe, a technique we've been using since antiquity to make spectacles. However, these simple lenses also had serious image defects. It was difficult for early astronomers like Galileo to find clear, uniform glass for their telescopes. Most glass lenses contained little bubbles and iron impurities that gave them a greenish hue, an issue that haunted telescope-lens makers for decades. To rid our view of these aberrations, Sir Isaac Newton swapped the lens in 1668 for a metal mirror. And yet, it was still a crude version that tarnished easily and suffered from low reflectivity. More than a hundred years passed before astronomer William Herschel built the then-largest telescope, at forty feet long, to bring the cosmic distance closer to our star-hungry eyes.

Astronomers have long recognized that another complication exists. We can only see cosmic explosions that reach Earth at a certain intensity of light. This finally changed with the invention of astrophotography in the nineteenth century. Cameras attached to telescopes achieve what our eyes alone fail to do: they can remain wide open, make long exposures, and create a permanent photographic archive. Our first photograph of outer space, snapped in 1840, was of the cold desert of the moon, its barren landscape a remnant of its volcanic history, and its pockmarked face the aftermath of meteorite collisions. It wasn't until the end of the century that detailed starry nights were recorded on photographic plates, the spangled heavens fixed in place like dark specks of dust to glue.

Another breakthrough came in 1857 when Karl August von Steinheil and Léon Foucault discovered a way to chemically deposit silver on glass to make a larger, more stable reflective mirror. In the years that followed, the telescope continued to grow larger and more sophisticated, with giant mirrors taking the place of the original lenses. When it comes to telescope mirrors, tiny doesn't seem as if it should be the name of the game. Bigger should be better, right? And for many years that was the case. The bigger the mirror, after all, the greater its power to catch light. However, big mirrors are typically heavy mirrors, and heavy mirrors store more heat—which is *not* good. Most telescopes in the 1970s and '80s, when Angel was hard at work on this issue, didn't live up to their potential because of their size. At twelve feet in diameter, mirrors are already large enough to hold heat that spoils their images, he explains. Not only that, but the larger the mirror, the longer the distance from the mirror to the focal point, and thus the larger the dome needed to house the telescope. A twelve-foot mirror needs "a cathedral-sized dome" to focus the light of distant stars.

Angel scooted around this issue by designing a honeycomb mirror that can support the mirror's face and achieve high optical quality. The large mirror is broken up into smaller honeycomb-shaped mirrors like the pattern found in beehives, so each mirror fits perfectly with the next one, creating a roughly circular overall mirror. His honeycomb design weighs about a fifth as much as the solid mirrors. As Angel explained to the National Inventors Hall of Fame, we can find in nature "both the fundamentals of how the universe works but also in engineering, how to build stuff. Nature has evolved honeycomb structures to make stiff, lightweight things, and that's very useful in a telescope mirror."[20]

The next great leap in astronomy took the honeycomb mirror and launched it into low-Earth orbit, unshackling our view of the cosmos from the confines of our planet's atmospheric distortion. The Hubble Space Telescope, the granddaddy of telescopes, has a primary mirror with a honeycomb core, reducing its weight from 8,000 pounds (3,630 kilograms) to 1,800 pounds (815 kilograms). Hubble works so well that its ability to capture faint light is the equivalent of a human eye in New York seeing the glow of a pair of fireflies in Tokyo.

Of course Hubble isn't the only star-seer floating in the unfathomable cold of space. Dozens have left the confines of Earth for the vacuum of the cosmos, from the Kepler

Space Telescope, designed to hunt out Earth-sized planets whizzing around nuclear furnaces similar to our sun, to the Neil Gehrels Swift Observatory, which detects the most energetic explosions in the universe, known as gamma-ray bursts. Each one provides a little slice of heaven, a glance into an uncharted realm. They show us all kinds of things. What they can't show, however, is X-rays.

The Starry Messenger

THE X-RAY TELESCOPE has had a slow, difficult birth. To understand its delivery into astronomy, I called Judith Racusin, an astrophysicist at NASA's Goddard Space Flight Center in Maryland. Racusin studies X-ray and gamma-ray observations of the most energetic electromagnetic events in the universe. It's fitting that a scientist who observes the energetic goings-on in space speaks with such pizzazz, her gestures ranging from the sweeping to small, precise motions.

Most of us are familiar with X-rays, though our first thought when we hear of them is likely of hospitals. However, X-rays are also belched out from celestial bodies that are searingly hot, such as fast-moving material swirling into black holes, galaxy clusters, exploding stars, and other high-energy events. It's tricky to detect X-rays because their intense energy zooms through everyday objects, hence their use in medical devices and airport scanners. One X-ray photon is hundreds to thousands of times more energetic than an optical photon (those within our range of vision). In fact, X-ray photons are so powerful they beam right through typical telescopes. So why are astronomers particularly eager to observe them?

X-rays give us a more complete view of the evolution of stars and galaxies. The Hubble Space Telescope only observes the cosmos in near-infrared, optical, and ultraviolet radiation, while the Spitzer Space Telescope detects infrared radiation. Until recently, there has been a gap in our knowledge of cosmic X-rays, which can reveal some of the most extreme physics in the universe. Some stars, like our sun, become white dwarfs at the end of their lives, exhausting all their nuclear fuel and becoming a dense core. Other doomed stars that are at least five times larger than our sun go out with catastrophic force, collapsing to form a neutron star or black hole and blasting most of themselves into space in one of the most violent events in nature, known as a supernova. NASA gives the mind-bending example of imagining an object 1 million times the mass of Earth collapsing in on itself in fifteen seconds, creating enormous shock waves and a violent light show. If a supernova happened within a dozen or so light-years of Earth's cosmic backyard, all life would be vaporized and extinguished. Luckily, our sun isn't fated to such a demise.

Whether supernovas have exterminated other planets is not known, but what we do know is that supernovas spread life-giving elements such as carbon, nitrogen, oxygen, silicon, calcium, and iron over thousands of light-years. If a supernova shock wave whams into a cloud of dust and gas, it can trigger the gas to collapse and, after a million or so years, create a new generation of stars that shine as dazzling beacons in the cosmic darkness. All the heavy elements in the universe except hydrogen and helium were formed in the middle of the massive stars. If you search for "supernova photos" in NASA's archives, you'll find explosions that look like prismatic gemstones, the colors revealing all we cannot see with unaided eyes. Their destruction is imbued with an almost ethereal beauty in the black expanse. When stars die, their metallic shrapnel explodes into the cosmos, and those elements go on to form the next generation of stars, planets, and perhaps, eventually, even you and me; our bodies are composed of elements born from massive stars. We survive because we are cradled by circumstance, time, and a collision of mysteries—the consequences of our actions rippling across time and space like gravitational waves.

Both fortunately and unfortunately, ground-based instruments can't capture X-rays, Racusin tells me, because their electromagnetic waves can't pierce the halo of Earth's atmosphere. This skin of gases is our translucent security blanket, protecting life as we know it. The atmosphere harbors the oxygen we breathe, shields us from harmful solar radiation, and destroys small rogue meteors that slam into it from the nether regions of the universe. Yet its protection also distorts our view of the cosmos.

When starlight streams through the universe, the data is delivered to our planet's atmospheric doorstep. But in the final moment before the data reaches our eyes, it gets mangled by our protective shield. The twinkling stars we see are more than a cute image in a nursery rhyme; that twinkling is a distortion created by atmospheric turbulence, which bends the light and jiggles the beam before we can properly observe it. In order to get the best image we can, we construct telescopes on mountaintops and in regions where there is less turbulence, but the distortion remains. The strategy for the twenty-first century, then, is to fly outside the boundaries of Earth's atmosphere and use telescopes in space.

And yet, to observe the X-ray universe, astronomers needed something utterly new. Angel was one of the first to design a telescope that would get the job done. In 1978, at forty-four years old, Angel was on a plane flying home from a Christmas holiday in England, passing the time by flipping through the pages of the December issue of *Scientific American*. An article titled "Animal Eyes with Mirror Optics" caught his attention. In the depths of Earth's oceans—a place less explored by humans than the surface of the moon—was a creature with eyes unlike any Angel had ever seen: the lobster. The lobster's inky orbs forgo lenses like our own and instead harness an entirely different technique to focus light.

It was then and there—somewhere over the Atlantic Ocean—that the lowly lobster crawled out of the ocean and into the pages of astronomy, providing Angel with the inspiration for a telescope that could catch X-rays millions of light-years away. His idea preceded Hubble by a decade, but it was only many decades later that our technology was finally sophisticated enough for his vision to launch into space.

Eye of the Beholder

ON A CRISP day in October, a friend and I are hiking a trail near the Pacific coast, talking about all sorts of things: work, future plans, and...

"Lobsters?" My friend wrinkles her nose in disgust. "What about them?"

She continues her rant as we make our way up a steep trail path: "They're red [not when they are alive], they scream when you cook them [they don't have vocal cords], and they have one weird, massive claw [that's true]. Why would you want to study them?"

To be fair to my friend, it's easy to judge a creature that urinates out of its face. You'll read in textbooks that lobsters have brains the size of a pen tip and poor eyesight (some scientists don't even think they have a brain, just a collection of nerve endings called ganglia), but such dismissals do them a disservice. The lobster eye is 256 times more powerful at catching beams of light in the dark than a human eye is in daylight. Lobsters use this skill to see in murky waters where they search for food. Our vision would be useless there.

The man who wrote the article that inspired Angel on the plane was none other than Michael Land, a British neurobiologist at the Sussex Centre for Neuroscience. Land spent decades probing the vision of jumping spiders, mosquitoes, clams, mantis shrimp, and, of course, lobsters, leading Richard Dawkins to dub him "the King Midas of animal eye research."[21]

"Evolution has exploited nearly every optical principle known to physics," writes Land. Evolution has produced pit eyes, pinhole eyes, compound eyes, and lensed eyes. Two eyes are a common sight, but not the only option: ogre-faced spiders have eight eyes, chitons have eyes speckled all over their backs, starfish have eyes on the tips of their arms, and dragonflies have 28,000 lenses. Some bacteria, such as single-celled pond slime, can sense light by using their bodies like mini-lenses. When light hits the curved surface of their cell, it refracts to a point on the other side and triggers them to pull themselves with little tentacles toward the source of light.

As we hiked up the trail's winding switchbacks, I set myself the task of noticing all the eyes around us. There were tawny sparrows hopping about and throwing side-glances in our direction; a bushy-tailed squirrel scuttled up a tree and poked its head out from behind a notched trunk, spying us from above with dark obsidian eyes. There was a deer, whose body froze, its unblinking eyes focused and resolute. There were even the tiny eyes of a rattlesnake that whipped its body into the brambles, out of sight as quick as we saw it. And of course, there were the fidgety eyes of humans, too, the ones that acknowledge our presence and then veer to the side, not lingering too long on the contours of our faces like a child's or a lover's would.

The world, it seems, is crowded with eyes. This unnerving realization is tempered by the knowledge that most of them cannot see us, at least not in the way we view the world. Many of these eyes can distinguish only light from dark, others just movement. Why did Earth invent such visual diversity? The answer is simple enough even if the visual hardware isn't: Eyes exist because the sun does. The eye is a device to gather light from the sun's rays and gain information about our environment.

The development of lenses in the eye, including our own, was a milestone event in evolution, emerging at the end of the Devonian period, some 360 million years ago. A clear, flexible jelly grew over the pinhole to help focus light and, in doing so, introduced entirely new ways of seeing (we owe a lot in life to variations of goo). To focus light onto a single point, the lens in our eye is thicker in the middle and thins out toward the edges. Muscles in the eye contract when looking at near objects, making the lens thicker. They relax when looking far away, with the lens becoming thinner and flatter to change the focus of the light. All of this information is then projected into our brain for it to decipher. This is an incredibly common design for the eye, but it's not the only one. There is a rarer design, one that scientists once believed was a "biological impossibility."[22]

We're talking here about the mirrors found in the eyes of lobsters and scallops. Up close, the eye of a lobster is curved like the dome of an observatory, "but under the microscope a lobster's eye looks like perfect graph paper," Angel told *Science* magazine.[23] That eye is composed of millions of tiny mirrored tubes, or "micro-channels," each measuring about 20 microns across (a micron is one-millionth of a meter, or twenty times smaller than the period at the end of this sentence). These mirrored tubes collect as much light as possible from all angles. Light hits the smooth surface of the mirrors and is reflected onto a single point on the retina. Where humans see by refraction—a process that flips the world for our brains to see—lobsters see by reflection, no flipping necessary.

The lobster's eye offers an unusual example of right angles in nature. Look at a forest, a bumblebee, the swirl of an ear, a seed, an octopus—no right angles. If we look at humanity's designs—desks, street corners, sheets of paper, floor tiles—right angles are everywhere. They're so ubiquitous we scarcely notice them in our windows, rugs, or packaging boxes. Where humans build with right angles in mind, nature likes her curves. Scientists thought mirrors in eyes were out of the question because, for one, mirrors are made with polished metal, and creatures don't make metal for reflective surfaces. Second, rectangles are seldom seen in nature. Yet one only has to think of a butterfly's shiny wings to know that nature does make iridescence. A common way to shimmer in nature makes use of cytoplasm and guanine crystals, found in fish scales and scallop eyes. Guanine is an essential component of DNA and RNA, and its crystalline form is one of the most widespread organic crystals in nature. Creatures can use guanine crystals to manipulate light, create color, or improve their eyesight. From an evolutionary perspective, the convergence of guanine solutions in the wild probably comes from the fact that guanine crystals have one of the highest known refractive indices (a measure of how slow or fast light travels through a material) for biological materials. In lobsters, the tiles of guanine crystals are stacked twenty to thirty sheets thick and separated by layers of cytoplasm.

In the 1940s, instead of the traditional method of making a mirror—with a single layer of silver or aluminum—mirror makers began layering very thin films of alternating high- and low-refractive index. "This turns out to be the way living organisms have made mirrors all along!" wrote Land in *Scientific American*.[24] The lobster was one of the first creatures we discovered with this visual hardware, but since then we've found mirrored eyes in scallops and dragonfish too. One creature, the brownsnout spookfish, is a visual overachiever with both a lensed pair of eyes that point upward and a mirrored pair of eyes that point downward. Other visual oddities include cubozoans, venomous marine creatures similar to jellyfish but faster. They have sixteen crude, light-sensitive depressions (half slits, half pits) and eight sophisticated eyes with a lens, retina, and cornea each, which is truly bizarre since cubozoans don't have a brain capable of processing the images. Why have these visual tools then? The answer is unclear.

The architecture of a lobster eye makes more sense, although it's certainly an anomaly. Typically, animals like insects have hexagonal eye structures because this provides the greatest packing density. The lobster's eyes are instead rectangles, in which the length of the mirrored tubes is double or triple the width. This ensures the rays are reflected off two of the mirrored faces but not more. The mirrors also form corner reflectors, meaning light returns in the direction from which it came. This sometimes happens in clothing shops or hair salons, where no matter what direction you look when you peer into the mirror, it reflects your image back. Angel found this intriguing. Since X-rays can power right through technology, they need to be snagged at a slight angle—or grazing incidence. The lobster eye provides a beautiful template for skidding the X-rays off the mirrored walls and onto a single point. "I had my article half written by the time we landed in Chicago," said Angel.[25] The title of that article said it all: "Lobster Eyes as X-ray Telescopes."

But to actually make it happen, Angel had to create the smoothest mirror ever built for a telescope.

Unexplored Frontiers

ANGEL'S LOBSTER-INSPIRED DESIGN was ahead of its time. The primary stumbling block was the micro size of the reflective tubes and how to manufacture mirrors at such a small scale.

"Everyone thought it's a good idea but it's challenging technologically because it's kind of messy to get the light to focus neatly," says Racusin. "The coatings on those mirrors have to be really smooth. If it's bumpy inside your microscopic grid, your photons are going to bounce all over the place and not focus cleanly onto your focal plane."[26]

For pretty much any telescope, there's a trade-off between how much of the cosmos you can scan and how sensitive the instrument is. The most sensitive telescopes are those that look at the tiniest parts of the sky. The Hubble Space Telescope, for example, is supersensitive with a tiny field of view. When Angel proposed a lobster-inspired telescope, it was to survey signals from the universe as an all-sky monitor. Just like lobsters have a 180-degree view to assess their muddy habitat, Angel's telescope would scan the cosmos in 180 degrees for flashes of X-rays.

It wasn't until 2001, twenty-three years after Angel came up with the lobster-inspired telescope, that a team led by the late George Fraser at the University of Leicester took Angel's vision and made it a reality, pioneering the development of a revolutionary new X-ray telescope. Fraser's enthusiasm for a new age of X-ray astronomy is crystal clear: "The scientific impact of Lobster will span all of astronomy—from studies of the X-ray emission of comets to stars and quasars, from regular X-ray binaries to the catastrophic events of supernova and the enigmatic gamma-ray bursts."[27]

Fraser's team at the University of Leicester and the European Space Agency partnered with Photonis, the world's only manufacturer of a forty-step glass formula for making square-pore micro-channel plates. Traditionally, shrinking technology like a mirror is a greater challenge than making it bigger. To mimic the mini-mirrors in a lobster's eye, a square block of glass is first heated and stretched by a weight. This step is repeated over and over again to make the glass smaller and smaller. The block is actually made up of two types of glass with different composition: one on the inside, and one on the exterior. When the glass is stretched to 20 microns, a chemical dissolves the internal glass, leaving only the external glass, now a tube just a few microns in thickness. These micro-channels are then assembled into bundles of twenty-five and cut along their width at a ninety-degree angle to create a wafer. The bundle of micro-channels is heated again and pressed to transform a flat material into a curved one. Finally, the micro-channels are dunked in an iridium bath to increase their X-ray reflectivity. If no complications arise and no small defects form, the entire process takes around six months and produces twenty-one wafers. This may seem like a niche technology, but it translates to other sectors too: Photonis was selected by the French army to supply state-of-the-art night-vision goggles. The company also creates low-light cameras used by fish farmers to inspect their nets and fish health.

With Photonis's help, it was finally possible to make Angel's vision of X-ray telescopes a reality. In 2018, a joint mission with the European Space Agency and the Japan Aerospace Exploration Agency, called BepiColombo, was launched on a seven-year journey to Mercury. On board is the Mercury Imaging X-ray Spectrometer (MIXS). MIXS looks just how you'd picture a black, 3.25-foot (1-meter) telescope to look. Its insides, however, feature more than a thousand tiny holes, each made of mirrored walls designed to focus X-rays onto a detector. An aluminum film on the front of the mirrors prevents stray light from washing out the image, while still allowing X-rays to stream through. The University of Leicester says the mission "features the first true imaging X-ray telescope to be used in planetary science."[28]

BepiColombo's goal is to learn about Mercury's formation. Now that Pluto has been demoted, Mercury is the smallest planet in the solar system (it's about the size of the continental United States) and the closest to the sun. As such, it is a planet of extremes. Upon arrival, the spacecraft will endure a blistering 662°F (350°C); the craft will orbit, not land, because Mercury's surface is hot enough to melt the trinkets of our space machines. Temperatures there can climb to 800°F (427°C) during the day and, with no atmosphere to keep in the heat, dip to −290°F (−179°C) at night.

A second lobster-inspired X-ray imaging telescope (called MXS) weighs just 2.2 pounds (1 kilogram) and will be launched aboard the Space Variable Objects Monitor (SVOM), a joint Chinese-French satellite mission with a projected date of June 2022. Its incredibly small size is in large part due to Photonis's glass manufacturing, a useful technique for missions where minimal weight and size are precious metrics. The SVOM observatory will explore gamma rays, which are explosive embers fizzling out in the black velvet of space. Gamma-ray bursts are a trillion times more powerful than visible light and emit more energy than the sun ever will during its 10-billion-year life. Despite the power of gamma-ray bursts, they were only discovered in the 1960s in a serendipitous turn of events.

The possibility of nuclear warfare during the Cold War spurred the U.S. military to launch spy satellites into space to spot signs of Soviet nuclear tests. They were on the lookout for gamma rays, which stream from nuclear explosions, lightning, and radioactive decay. The military didn't uncover any signs from the Soviet Union, but they did receive enigmatic signals from outer space. For thirty years, these signals were a mystery—until scientists finally pinpointed the source as a galaxy 6 billion light-years away.

There are two types of gamma-ray bursts: short and long. Their creation stories differ but their demise is the same: a black hole, a region of space where gravity is so strong that nothing, not even light, can escape. Short gamma-ray bursts linger for less than two seconds and are likely caused by two neutron stars crashing into each other and forming a black hole. A long burst endures for greater than two seconds and is caused by the dramatic death of stars imploding to become a black hole. The MXS telescope will capture not the explosion itself but its afterglow, the fading remnants that remain observable at less energetic wavelengths.

As NASA says, "If we could see gamma rays, the night sky would look strange and unfamiliar. The familiar view of constantly shining constellations would be replaced by ever-changing bursts of high-energy gamma radiation that last fractions of a second to minutes, popping like cosmic flashbulbs, momentarily dominating the gamma-ray sky and then fading."[29]

Racusin is keen to explore SVOM's use as a sky monitor. The idea is that a wide X-ray telescope says, "I have detected something interesting over here," and a more sensitive telescope follows up with observations within about a minute. This all happens autonomously. It is teamwork at its cosmic finest.

The cosmos captures our imagination like nothing else because it contains the story of how we came to be—it is Earth's birthplace. To our ancestors just a few generations ago, we would seem to harness the power of gods with our technology. We have launched spacecraft into orbit using millions of pounds of thrust; we have stepped on the moon, we've tamed the energy of the sun with solar panels, and we have a constellation of satellites in space to beam signals to our homes. And to think: all of these innovations were devised within the last 250 years, a throng of technological triumphs that didn't exist in the minds of humanity except perhaps as reveries.

The animal kingdom is rife with other biological sensors, too. There is the electroreception of a platypus, the infrared detection of snakes, the polarized vision of octopuses, the fire detection of jewel beetles, and the magnetic sensor of bees, to name but a few. Sharks are one of the best biological conductors of electricity, using a network of jelly-filled pores around their face to sense differences in the electrical charge of an animal and the water around them. Pit vipers detect infrared light with "night-vision goggles" sensitive enough to notice when temperatures vary a thousandth of a degree. Roundworms use a single nerve to detect Earth's magnetic field. And some animals navigate by the stars, similar to our ancestors. The dung beetle steers by light from the Milky Way on moonless nights, a luminous smear across the dark canvas of the sky. Scientists figured this out by using LEDs to construct a fake night sky with a Milky Way. When given a random pattern of pinprick stars, the dung beetles were utterly lost. Only when the Milky Way was displayed, with its varied brightness in certain regions, did they figure out which way to go. Could we program robots to do the same, they wondered?

Most of us can no longer navigate by the radiance of stars like our nomad forebears once did; our memory of cosmic patterns is fading. Yet in other ways, we are more connected to the cosmos than ever before. Astronomers have moved from staring through telescopes with furrowed brows to fashioning celestial Goliaths that can send us images, texts, and data from the depths of space. Even when we can't physically get our bodies somewhere, we can send robots as our eyes and barometers, mini-scientists punching holes into another planet and collecting samples. The lobster is a reminder that Earth's creatures are portals into other ways of seeing the world; its mirrored eyes now helping us to voyage beyond our own sight and explore the cosmic elements that led to the creation of life on Earth.

As writer Nan Shepherd says, we wander the world with "gullible eyes."[30] By closely observing the world around us, we can come to understand the limitations of our own ways of seeing and building. Perhaps, as poet Diane Ackerman writes, "first we need to see ourselves from different angles, in many mirrors, as a very young species, both blessed and cursed by our prowess. Instead of ignoring or plundering nature, we need to refine our natural place in it."[31]

3

DRINKING FROM A CLOUD

Coastal Redwoods Inspire Fog-Catching Harps

"The General Assembly... recognizes the right to safe and clean drinking water and sanitation as a human right that is essential for the full enjoyment of life and all human rights."

UN GENERAL ASSEMBLY RESOLUTION 64/292 (2010)

IN THE DRY Andean foothills on the outskirts of Lima, Peru, lives a mother of three in a village with no running water. She washes potatoes in a dirty bucket. She uses a hole in the ground as a toilet and a cement tub to bathe her children in cold water. On this day, she picks up her youngest, his rosy cheeks smudged with dirt, and puts him in the tub. Cool water trickles over his dark hair; he squeezes his brown eyes shut as it streams over his bare shoulders. Finding water to bathe in is not a problem—at least not compared with finding clean drinking water. It is dry here; the yearly rainfall is just one inch. And so this mother, and others like her, must rely on the trucks that grumble through the village to bring water. It's always expensive, and there's no guarantee that it's actually safe to drink.

The mother's story is typical of some 40 percent of the world's population. Women in rural southwest Morocco, for example, wake up as early as 4:00 AM and spend up to three and a half hours a day fetching water for their families. Sometimes they arrive home with barren buckets, the wells having dried up in the summer heat. They are known as "water guardians," but there is nothing glamorous about their sacrifice. Young girls often forfeit school to help their mothers, perpetuating a cycle of poverty.

If we zoom away from these women, away from the continent and Earth entirely, and view our planet from space, it's easy to wonder what all the fuss is about. Three-quarters of the planet is covered with water, after all. Surely there is enough for everyone, no? Well... no. This vision, tantalizing as it appears, is misleading. Most of that water is salty and unfit for consumption; only 3 percent is fresh water, and two-thirds of this is frozen in glaciers or polar ice. And so, when it comes to drinkable water, we are left with just 1 percent of what the planet holds, found in snow, streams, and lakes. The grievance of the ancient mariner rings loud and clear: "Water, water, everywhere, not any drop to drink."

According to the United Nations, about 4 billion people experience severe water scarcity during at least one month of the year, and 2.2 billion lack access to safely managed drinking water services. These challenges contribute to a startling statistic: 80 percent of illnesses in developing countries are linked to unsafe water. Incredibly, the sanitation system of ancient Rome would be an improvement for nearly 3 billion people today. Access to clean drinking water has become one of the greatest humanitarian challenges of the twenty-first century. Lest we forget, we need water not just to drink and bathe but to eat; most of our water use is for agriculture and industry. Efforts to convert salt water to fresh water have made desalination a key resource, but it is energy-intensive and expensive. The disposal of salt that's removed from the water is another concern.

There is potentially another freshwater source we can tap. My journey to find that water took me on a six-hour drive north of San Francisco to the largest contiguous old-growth coast redwood forest in the world.

Home of the Fog Giants

UP THE COAST from an autumn day in San Francisco, the streets get windier and the fog rolls in. I'm on my way to meet naturalist John "Griff" Griffith in Humboldt Redwoods State Park, located along the Eel River, California. The region is known for a plant that you can catch a whiff of as you drive through town. Humboldt's cannabis industry is among the finest and most fertile in the county. It was the lure of gold and timber that initially brought people to this land, but for the last four decades it's the cannabis that's made them stay. The town's real pride and joy, however, tower over those plants like sentinels.

My visit was kindled by my conversation the week before with Brook Kennedy, a professor of design at Virginia Polytechnic Institute with a soft, melodic voice. During a jog one morning, Kennedy's mind wandered back a few years to a time when he roamed "the home of the fog giants," where coast redwoods shoot up hundreds of feet from the soil.

Here, his idea for how to help people who live in water-stretched regions of the world took root and inspired him on a journey that would land him in the pages of top news outlets around the world. To appreciate his invention, and the seed of its inspiration, I met Griff at the state park visitors' center.

A seventh-generation Californian with a bushy goatee, Griff treks through the redwoods in khaki pants and a double-chest-pocket shirt, wearing a wide-brimmed naturalist's hat and a name tag sewn to his breast pocket. It's a uniform he waited more than twenty-seven years to don, after swooping into the modest town every year to ask about the position. Finally, in January 2020, the position opened up, and Griff was just the right person for the job.

As we hike deep into the heart of the redwoods, the snap, crackle, pop of twigs underfoot add a musicality to our chat. In his youth, Griff protected the coast redwoods as part of the "Redwood Wars," in which protesters did everything they could to defend the trees from being felled by Pacific Lumber. An activist died by a fallen tree, a bomb was planted under another's car, and tree-sitters were wrestled from branches by county sheriff deputies. Griff supplied the tree-sitters with food, which they swapped for jars of feces that he carried out. Griff saw himself as an ecological protector. Nowadays, his activist passion is directed toward nature education and toward highlighting the work of Laura Perrott Mahan, a woman who, in the 1920s, galvanized early support to save the redwoods, but whose gender slipped her behind the men who later founded the park with their money.

Redwoods are some of the oldest, tallest trees in the world; a giant named Hyperion soars 380 feet from the ground, taller than the Statue of Liberty. Their trunks can grow as large as twenty-four feet in diameter—wide enough that, decades ago, people carved tunnels into some of them for cars to drive through. The genesis of these giants is a seed the size of a flake of oatmeal, which stores all that the tree needs to interact with its environment for 2,000 years, the equivalent of eighty generations of humans. The current redwoods were alive when tea was first cultivated in China, when the decimal system was formulated in Crete, and when Jesus was born in Christianity. Their wood contains an abundance of tannins, which gives them their ruddy color and protects them from slow, colorless assassins: fungi. Trees, like us, are susceptible to fungal infections and in some cases can even transfer them to humans.

All of this and more makes it hard to tell a story about redwoods that hasn't already been told. There are tales about the world's tallest tree in *National Geographic*—complete with a fold-out spread to highlight its height—and stories about their thousand-year life spans and their destruction by European settlers in the Americas. There are stories about albino redwoods that spread their branches like ghostly fingers, and articles about the psychology of humans amid such grandeur. There are even quests for the undisclosed location of that tree from the *National Geographic* spread. But it's possible that the fog droplets that collect on these titans have been overlooked by journalists and hikers, brushed aside by those on another journey. Yet if we look again, these droplets hold prisms of potential in their spheres, a small world of vital proportions in the richly textured forest growth.

The deeper Griff and I trek into the thick of the trees, the darker and damper our world becomes, and the less noise we make on the forest floor. Five-fingered ferns splay their green leaves, and mousy birds scamper up the sides of fallen trees, the *scht-scht-scht* of their claws against the bark breaking the silence. As fog creeps in, the air cools and the trees fade in a wispy gray veil. We are walking in a cloud.

This isn't an exaggeration. Fog is simply a cloud low to the ground, and redwoods use it to make rain on a foggy day, saturating the forest floor with the same amount of water as a storm. When I first contacted one of the foremost experts in redwoods and coastal fog, Todd Dawson, he was busy helping with the wildfires ravaging California. Later we finally got a chance to chat, and he told me that redwoods are "like sprinklers, if you will, that are irrigating the entire understory plant community as well as the redwoods themselves."[32]

This fog, like all water, is a bit sticky. If water wasn't sticky, we wouldn't see drops cling to our car windows during a rainstorm or stick to our skin when we step out of the shower. The tree's needles take advantage of this and grab the fog, guiding the airborne water to the soil below.

Other plants such as ferns, shrubs, and moss—collectively called epiphytes—grow on the redwoods' branches like hairy green limbs. As much as 1,600 pounds (725 kilograms) of epiphytes can grow on a single redwood. The plants are not parasitic but rather signs of a healthy ecosystem. Redwoods support a teeming community in their canopies. To study these lofty ecosystems, researchers shoot a crossbow with an arrow threaded with a fishing line over a branch, sometimes over two hundred feet (60 meters) high, and then haul up a sturdier rope. They loop the rope through a harness on their waist, tie themselves in, and climb up the tree using mechanical ascenders in each hand, which clamp to the rope and tighten when weighted with their body. As the researchers push off a foot loop, the weight on the ascenders loosens, allowing them to thrust the ascenders up the rope. They do this again and again, inchworming their way up the tree (climbers call it jugging). Once in the thick of the branches, they toss weighted ropes over higher and higher limbs until they reach the tree's crown. Here they can find wandering salamanders nestled in limb crotches and the endangered marbled murrelets, seabirds that only nest high in the branches of old cone-bearing trees. These branches are wide enough to prevent the seabird's single speckled egg from rolling off hundreds of feet to the ground. Occasionally, a rare ring-tailed cat also slinks in and makes an appearance. In 2016, a new species of lichen, dubbed "redwood stubble," was found sprouting from the trees' bark. Then there are the red tree voles that live nearly their entire lives in the old-growth treetops; as the cold fog drifts across the branches, the red tree voles lick the drops clinging to the needles.

Though clouds may seem weightless when we're lying on our backs in the grass, finding shapes in their contours, nothing could be further from the truth. A puffy cloud floating in from the coast can weigh up to 1 million pounds and carry billions and billions of water droplets. The cloud we see is actually a lot larger than we think it is, roughly 0.6 miles (1 kilometer) on all sides if molded into the shape of a box rather than an amorphous blob. The clouds shapeshifting high above our heads float because the density of the dry air below them is greater than the density of the water droplets spread out over such a large distance.

Where most of us lament fog for slowing us down, delaying flights from taking off, or, as is often the case in San Francisco, shrouding sapphire skies into gray days, Professor Kennedy is fascinated by fog. One-third of the coastal redwood trees' water demand, Kennedy knew, is met by the fog that moves in off the Pacific. *Can humans capture water from fog too?* he wondered. Fog is present in nearly every country on Earth, and the idea has enormous potential from a humanitarian perspective, especially for some of the most water-starved regions on Earth. For centuries, engineers have tinkered with water systems, from water wells in the Neolithic era to lead pipes in Pompeii. In some ways, it should come as no surprise that we've turned to the idea of fog harvesting to fight water shortages. After all, we harvest many things: grain, grapes, fish, logs, energy, salt from seawater, and even cadavers. Why not fog? Earth's atmosphere carries 0.001 percent of all the water on our planet in the form of mist, fog, and vapor. It's easy to dismiss fog harvesting as an idea too small to make an impact, to shrug and say, "That won't fix everything." However, it's becoming increasingly apparent that solving our planet's water woes will require a tapestry of solutions. Fog harvesting might not work in, say, a large city in Texas, but for a small community in a desert region, it could tip the balance toward a better life.

The first recording of fog harvesting was in 1969 at Mpumalanga, South Africa, as a source of water for an air force base. A typical fog net design features flat, rectangular mesh stretched between two poles. Picture the mesh on a screen door and you'll be on the right track. As the wind passes through the net, droplets of fog get caught on the wires, like a swarm of flies in a webbing of sticky tape.

When the droplets swell to a large enough size, gravity does them in and they trickle down to a trough at the base of the contraption. The most well-known fog nets are FogQuest's harvesters, installed in the late 1980s in the remote fishing village of Chungungo near the Atacama Desert in Chile, the driest place on Earth. FogQuest collects drinking water from a coastal fog called *camanchaca* that drifts in as the people go about their day; the droplets flow into pipes and are stored in tanks, cisterns, or reservoirs. The harvesters worked so well for the village that nearly a hundred were installed. Together, they provided an average of 4,000 gallons (15,000 liters) of clean water each day, or about 13 gallons (50 liters) of water per person for a village of three hundred people. The endeavor was one of the first fog-harvesting projects on this scale in the world, and also one of the longest running. The system was in use for ten years, and during this time the population of Chungungo doubled in size from three hundred to six hundred residents. Cloud capture was a kind of salvation, rescuing them from trucks that brought water to the village, often with a high degree of contamination. At the turn of the twenty-first century, local politicians decided to stop repairing the fog harvesters and fund a new pipeline. By 2009, the pipeline had not been constructed and the fog harvesters had fallen into disrepair.

Still, excitement about water harvesting has come from all angles, including water-stretched South Africa and the United Arab Emirates, as well as cannabis farmers in Southern California. Peru is losing its glacial water and Chile has the Atacama Desert. Morocco has little water, but they do have fog. There's a need in Spain, Portugal, and parts of India and the Middle East. Oman is trying to reforest some of its desert.

"On every continent, there's an opportunity. Particularly in South America and in Africa," says Kennedy.[33] However, when he came up with his vision of harvesting fog, he needed more heads in the game, so he recruited a man who loves everything about water droplets. Dr. Jonathan Boreyko is a mechanical engineer and professor who directs Virginia Tech's Nature-Inspired Fluids and Interfaces Lab; the jumping-off point of Boreyko's lab is that nature has already solved some of the challenges humanity faces. The two professors nurture vastly different passions, but they unite on one idea: we can do better than traditional fog nets.

Innovation with fog nets has stagnated, says Boreyko. We need to ask more questions. Do thinner or thicker wires work better? What material does fog stick to best? Traditional nets suffers from a dual constraint: Too tight and the water droplets get clogged. Too loose and the fog passes through the net uncaptured. It's a conundrum that, on the surface, doesn't seem to leave much room for improvement. Kennedy and Boreyko chipped away at it, their minds slowly wrapping around another solution. In the end, their idea was as simple as it was profound, and took its cue from those redwoods that Kennedy had once found himself wondering about on a morning jog. Redwood needles don't look like tennis nets or screen door meshes; they are linear. And yet fog condenses on those needles, with water droplets swelling until they slide off, splash to the ground, and are soaked up by the trees' roots.

"The problem was not with the vertical wires," says Boreyko, "but the horizontal wires that cause droplets to get stuck when you're trying to drain them down."[34] The best way to imagine this process is to think of the water droplets as sprinters. With a traditional net, the sprinters have to jump a hurdle for every horizontal wire they encounter, and there are hundreds of hurdles to clear before they cross the finish line. To make it even more challenging, the water-drop sprinters must physically break off a tiny part of themselves each time they clear a hurdle—sort of like a track star that catches their foot every time they jump. This happens again and again as the drops continue down the net and finally plop into the collection container.

With this in mind, Boreyko and Kennedy designed a new kind of net. It no longer looked like the mesh on a screen door but instead resembled a musical harp, hence the name "fog harp." All that was left to do was test the thing. But first, says Boreyko, there's one additional hurdle that needs mentioning: "Almost everyone gets confused about the difference between condensation and fog. Even scientists in professional journals will often say 'dew' when they actually mean 'fog.'"

To get that straight, we need to back up a bit...

Luke Skywalker Was a Dew Harvester

"I THINK LUKE SKYWALKER in *Star Wars* was a dew harvester. His family had a moisture farm," says Boreyko, referring to the soon-to-be Jedi's farm located on the outskirts of the Jundland Wastes of Tatooine, a desert planet burdened by scorching suns. Water was a precious resource, so Luke's family used stationary condensation vaporators to draw water from the atmosphere.

Dew (or condensation) harvesters are not the same as fog harvesters. "Condensation harvesting" is a fancy way of saying you have a dehumidifier machine. Condensation harvesting is essentially a heat transfer process, and for it to work, warmer vapor in the air like humidity comes into contact with a surface that is kept constantly cold by electricity, turning the vapor into droplets of water. The process is similar to what happens on a cold water bottle on a hot day, drops of condensate collecting on the outside of the bottle. This kind of harvesting "is very, very energy-intensive," says Boreyko. "We know how to do it, but you have to have a tremendous amount of cooling to continuously turn vapor into liquid. It's astounding how much energy is required. It takes over 2,000 kilojoules of cooling energy just to turn 1 kilogram of water vapor into a liquid." That is enough energy to power more than sixty smartphones to full charge.

The beauty of fog harvesting is that it's the exact opposite. With a cloud, nature has already done the energy-intensive heat transfer. Water vapor forms around tiny particles floating in the air due to a temperature mismatch. The dehumidifying is done high in the sky, leaving only fog for us to collect. Scientists just need to find a way to capture it.

"Fog harvesting is much lazier than dew harvesting," says Boreyko. "You've let nature do all the work for you in terms of condensing liquid water, and all we're doing is catching it and drinking it."

But fog harvesting doesn't just come down to catching the fog drops—because with anything porous you put in the air, you're going to be snagging droplets. The key is figuring out that critical size where gravity wins, beats surface tension, and lets the droplet slide down into the collector. To test their fog harp design, Kennedy and Boreyko pitted it against traditional mesh in a Skywalkian battle of the harvesters.

Location: Kentland Farm in Blacksburg, Virginia.

First contender: a square mesh of stainless steel wire similar in diameter to the harp's wires. Result: The fog harp collected five to seventy-eight times more water, depending on the weather. It wasn't even a contest. Time for a new rival.

Second contender: coarser, thicker mesh than the first. Result: The fog harp collected water on only eight of its twenty-one days in the field, but it was still enough; its opponent collected water on just two days. The harp trickled to a win with twenty-one to twenty-five times more water collected.

Then, the real battle: the fog harp side by side with FogQuest's state-of-the-art Raschel mesh harvester. It was a fierce competition. In low winds, the FogQuest outperformed the fog harp. However, a summary of all days in the field saw the fog harp drip to first place. The test included a fair number of days with light fog and high wind—conditions in which the harp thrives. Light fog is the harp's bread and butter. Most fog nets need droplets to reach roughly 10 microliters in size before they plop into the collector. The harp only needs drops 0.01 microliters to slide them down the wires.

"We've basically got one hundred to even pushing one thousand times smaller volumes required for the droplets to be able to slide," says Boreyko. However, if there had been less wind, FogQuest likely would have won. This suggests the local conditions are a vital consideration before choosing a fog harvester.

The harp also excels in heavy fog because the water clogs traditional mesh nets. If the holes are plugged, wind doesn't pass through the net but instead whips around it. If this happens, the net doesn't catch any new droplets. "Whereas for the harp," says Boreyko, "it's anti-clogging because the droplets always slide at a smaller size than the holes between the wires."

Some scientists are using water-repellent surface coatings to make the drops slide down better, says Boreyko. And there is some success with that, but the coatings can possibly break down over time and get into the drinking water. Boreyko's philosophy is to use geometry instead of chemistry. "Why are we putting coatings on in the first place?" asks Boreyko. "Because the drops get stuck on the horizontal cross wires. If that's the case, then the geometrical solution is to take off all the cross wires."

Just like animals adapt to local niches, humans too could cater to the conditions in their own backyard. If there is a way to make a big impact on a local scale, it could have a profound, meaningful influence for some of the world's most impoverished people. Depending on the level of air pollution in the region, the water can be used either for drinking or for crops. And if drinking water is needed, purification techniques can be used.

In San Francisco, where there is usually fog, especially in areas like Noe Valley and Sunset District, "you could have a little herb garden that self-waters or small-scale agriculture for your home," says Kennedy. "The fog would drift in, and the harp would collect droplets and drip down into the soil of your little garden."

Another angle is to install fog harps on the top of Twin Peaks in San Francisco. "People could watch the fog roll in and the water drop off," says Kennedy. "And then fill up their little CamelBak or Hydro Flask, have a drink, and go on their merry way."

Boreyko adds: "Maybe in California where, if the water scarcity crisis is bad enough, a utility company could take charge of building a farm of harps. It's under the central control of the utility company. Then they could simply sell that water that is harvested as utility-grade water."

Fog harvesting is not a silver bullet, reminds Kennedy. "Fog harvesting is like part of a complete breakfast kind of thing. I don't think it's ever going to provide the kind of volume of water for the American lifestyle. It will provide clean water for your basic drinking or agriculture, or some sanitation needs."

The bigger issue, then, is how we can conserve water. California's Central Valley has been pumping water out of the ground at such an unsustainable rate that the land is literally sinking as a result (a process known as subsidence). In the last century, the valley has sunk 28 feet (8.5 meters). Other states are also suffering from subsidence, including south-central Arizona, where lands have sunk as much as 12.5 feet (3.8 meters) since the 1940s. Groundwater extraction has also caused significant water-level declines in Arkansas, Mississippi, Louisiana, and Tennessee. Countries like South Africa have managed to avoid completely running out of tap water mainly through conservation. Simple measures like reducing consumption can help.

"The point I'm trying to make," says Kennedy, "is that the fog harp is a brilliant technology for increasing water volume, but we can't look at it as a replacement for a lake. Hopefully, it will get people to think more about conservation as well. To think of water as a precious resource, which of course it is."

Animal Solutions

ALL ORGANISMS NEED water, salty or fresh, to survive. Water is the elixir of life, and it's no wonder that after 3.8 billion years of evolution, animals have developed some ingenious solutions for getting water from fog. The Namib Desert beetle in southern Africa lives in a sea of sand that stretches for almost 1,200 miles (2,000 kilometers). Here, if you're lucky, you'll catch a black dot scuttling up a sand dune, grains of sand tumbling behind its spindly legs. When the beetle reaches the dune's peak, it slows to a pause. There is no water in sight—yet. The beetle bows its head to the impending gift, its rear end rising to the sky in anticipation and exposing bumps peppered like poppy seeds on its back end. When the fog rolls in and drifts over the beetle's body, water droplets gather on its back. Scientists call the bumps *hydrophilic*, meaning "water-loving." The water swells and then rolls down the beetle's back and into its mouth. The beetle has inspired the creation of self-refilling water bottles with hydrophilic bumps that attract water straight from the air. Engineers are also investigating whether they can design tents that snag water droplets carried on the wind.

Then there's the African bush elephants that lack sweat glands so they flap their ears like fans and hold water in their wrinkled skin to keep cool, storing five to ten times more water than a smooth surface would. A barrel cactus in Mexico uses the tips of its red spines to collect pearls of water. Those pearls are then absorbed by the succulent in a process that's known as a Laplace pressure gradient (the difference in pressure between the inside and the outside of a curved surface). The cactus's pressure gradient moves fog droplets against gravity to the base of the spine, where it can be collected by the plant. The physics of the Laplace pressure gradient is complicated, but suffice it to say that scientists are exploring whether they can mimic the technique to create water collection devices.

And we can't forget the iconic Australian thorny devil, a lizard whose entire body bursts with spines. The thorny devil lives in dry sand country and moves like a grizzled old dinosaur, its mouth glowering in a moody pout. Grooves in its skin gather fresh water, which flows to its mouth against the force of gravity. This is achieved using capillary action, whereby water moves through narrow channels using adhesion, or "stickiness," and surface tension to overcome gravitational forces. To picture how this works, imagine dipping a piece of paper or a towel in water. Counterintuitively, the water travels up the paper, defying gravity.

This isn't the first instance of organisms using the stickiness of water to their advantage. The "lotus effect" is a prime example, lotus plants in this case making their leaves less sticky than water's natural stickiness. The plants use just the basic ingredients gifted to them by the environment to stay clean, no detergents needed. If you were to pour a glass of water on a tree leaf nearby, the liquid would smear across its green surface. Do the same with a lotus leaf and the water sticks together like a jiggly blob and rolls off the leaf, sweeping up dirt along the way like a broom. How do lotus plants do it? And why?

The why is pretty easy to explain: a clean leaf can absorb more sunlight than a leaf covered in mud. In a harsh world, these organisms need to arm themselves with the skills to survive and thrive. The ability to self-clean with no hands is a wild artistry that defied our imaginations for centuries—until we discovered its secret in nature's untamed lands. How lotus leaves tidy themselves is quite the spectacle, and it occurs at a super-small scale. The leaves are dappled with mini-bumps that are spiked with even tinier hairs. When water plops on the leaf, the drops are held aloft by the tiny hairs and buoyed by air pockets below, due to the irregular surface. This is important because water and air don't stick together as well as water and solids. This means the water is now more attracted to itself than the leaf, and it gathers into spheres of water that are more likely to roll off. To up the ante, the lotus's tiny bumps are coated with water-repellent wax. Together, the bumps and the wax make the leaf extra slippery, so that when the leaf is tilted at as little as five degrees, the water skates right off and with it dirt particles.

If this wasn't enough, Boreyko found that lotuses have evolved one more maneuver (these plants are serious masters of cleanliness). Where does the dew that slowly swells in the leaf's crevices at night go by morning? When the sun rises, the leaves are dry. There's no trace that a drop of water ever tainted its regal leaves.

"If you take a microscope and you watch the dew droplets as they grow on the surface," says Boreyko, a grin growing on his face, "the droplets will suddenly disappear from the microscope lens! They just disappear, and it's not due to gravity."

Subtle vibrations cause the dew to "launch out of the crevices with the acceleration of a spaceship," says Boreyko. These vibrations are easy to come by when you're a broad lotus leaf with a straw-thin stem.

If all this seems like a niche field of study, it isn't. The lotus's *superhydrophobicity*—or super water repellency—has already inspired everything from the self-cleaning glass on traffic control units, to stain-resistant textiles, to low-maintenance solar thermal energy collectors. It has also inspired water-free urinals and self-cleaning toilets to reduce disease transmission. Other prospects for water-phobic surfaces include vehicles, glass buildings, ships, and many more.

In drizzly Virginia, a state that feels a world away from the deserts of Namibia and the forests of California, Kennedy and Boreyko have a patent in hand for the fog harp and are in the commercialization stages of research and development. The first hurdle to any creation is the leap from mind to matter; the second is from physical matter to something that actually matters. To manufacture their fog harps on a mass scale, they need wires that are anti-tangling, a sturdy structure, and a price that is competitive. They don't want to make a pre-welded harp that's bulky, because that might prevent it from being carried by hand into villages. Kennedy is considering how to make them "knock-downable," to figure out how the parts can be packaged into small, flat packs and then unpacked and built like IKEA furniture.

That's one consideration; another is simplification. Is something fixable if it malfunctions? If no one in the community knows how to fix a high-tech but broken-down fog harvester, it becomes rubbish. "There have been plenty of companies that have gone off and tried to export, almost imperialistically, first-world tech to developing nations and what ends up happening is that it's a shiny new thing. Give it a couple of months, it breaks," says Kennedy. "No one locally knows how to repair it."

Kennedy also admits that there's a tension between making something that's beautiful but expensive and something that's cheap and not so beautiful but would still help people get water. Ultimately, it's not about a pretty thing. "Science is about people, not things," says Kennedy. "We can all pat ourselves on the back, like, 'Oh, look at our beautiful 3D-printed, robotically woven this or that,' but if it's not helping people, it's missing the point."

Back in Northern California, Griff and I hike under dozens of crooked branches fanned high above the understory. For these redwoods, the possibility of drinking from a cloud has ebbed somewhat. Fog frequency has decreased by one-third along the coast in the last sixty years. The remaining trees store 2,600 metric tons of carbon per hectare, according to a seven-year study. That is more than double the rate of the conifer trees of the Pacific Northwest or the eucalyptus forests of Australia. The redwoods store all that carbon in their trunks, root structures, and forest soil. In fact, coast redwood trees store more carbon per hectare than any other forest on Earth, including the Amazon rain forest. If you look at trees on an individual basis, redwoods are really impressive as a carbon sequestration organism. But as a forest, the coast redwoods' 2.2 million acres pale in comparison to the Amazon's more than 1.2 billion acres. Humans have now wiped out most of the coastal redwoods; only 5 percent remains.

Also overlooked are the coast redwood's chemicals. Scientists found that terpenes in the tree's needles not only give coniferous trees like coast redwoods their distinctive lemony fragrance, but also waft into the atmosphere and create particles that promote cloud formation. Terpenes interact with molecules in the air to form aerosols. As these aerosols rise in the atmosphere, they encounter water vapor and cause cloud droplets to form above forests. The clouds in turn reflect sunlight away from the Earth's surface and lower temperatures. And yet these days, ever-increasing levels of ammonia emissions from agriculture and sulfur dioxide from fossil fuels split the scented chemicals into aerosol particles that are too small to condense water vapor and trigger cloud formation. How much longer will these thousand-year giants last?

Griff launches himself over the back fence of the visitors' center using a tree as a foothold. When our journey down memory lane comes to an end, he unlocks the wood door to let me through the gate, back to the center. I return to the car for warmth as the evening fog drifts in with the autumn wind, leaves kicking up and dancing in the air; I spend the night sleeping in the back of the car under these old-growth monuments of the past.

When I finally return to San Francisco, I am met by a different fog, one the locals have fondly christened Karl over the years—a name for their complaints of another sunny day in California "ruined" by the fog.

4

WHO'S IN CHARGE?

Ants and Bees Inspire Efficient Routing Systems and Robotics

"So important are insects and other land-dwelling arthropods that if all were to disappear, humanity probably could not last more than a few months."

EDWARD O. WILSON, naturalist

BACK IN THE 1990S, Southwest Airlines had a problem. At Sky Harbor International Airport in Phoenix, some two hundred of its aircraft took off and landed each day on two runways, using gates at three concourses. All too often, those planes would land and then have to wait for gates to clear. An idle plane was a waste of time, energy, and money. After a long flight, passengers were jittery to get off the plane. The airline noted that even a fifteen-minute wait for an open gate would leave travelers visibly annoyed. How could they address the bottlenecks? Was there a way for pilots to choose the most efficient route to gates? The answers to these questions and others landed Southwest Airlines, of all places, among the miniature metropolis of millions of ants.

A Mother to All, a Queen to None

IN THE HOT, parched earth of the Chiricahua Mountains, where southern Arizona bleeds into New Mexico just off Route 533, Her Majesty the Queen delivers another egg deep in her lair. The swollen ant queen is no commander here; her nest is a delivery room, not an imperial command post. She mated mid-flight on her way to find a new home, where she will remain for the rest of her life. She used to have wings, but she ate them for sustenance. She doesn't need them anymore. She is queen of colony 154. Her sterile daughters, thousands of them, start the day at dawn, tendrils weaving out of the nest's hole and into the desert sun. Their existence depends on the seeds they forage, a trickle of early-day workers becoming a flock when food is found. By around noon, foraging and patrolling ceases, the heat too intense to continue any longer.

Zeep. An ant flies up as if by levitation. *Zeep.* A second, then a third, then a fourth, and then dozens. Biologist Deborah Gordon is kneeling in the dirt with an aspirator, socks over the cuffs of her pants, sucking up ants into a glass vial for research. For more than three decades, she has worked with red harvester ants at the Southwestern Research Station for Stanford University, and experience teaches you to cover up. An exposed ankle could mean a painful sting that blisters the skin, but really, this is a mild nuisance for Gordon; she's more interested in the ants' ability to operate as a colony without a central command.

"The basic mystery about ant colonies is that there is no management," writes Gordon in *Ants at Work*. "A functioning organization with no one in charge is so unlike the way humans operate as to be virtually inconceivable... No individual is aware of what must be done to complete any colony task."[35]

How do these bumbling critters that knock into each other as they forage for food get so much done without someone calling the shots? In large crowds, humans have been known to trample each other to death (think of the 2015 stampede that killed thousands around Mecca during the annual hajj pilgrimage). A single ant is not smart. A solitary ant will walk around in circles and die. Same with a hundred army ants. Yet gather a million of them and you have a "superorganism," creatures that perform complex tasks together as if they are of one mind. This is known as "collective intelligence." The study of insects, rather than a school of fish or a flock of birds, is called "swarm intelligence." It can look messy, tracking the moves of billions of ant acts across the vast network; in isolation, it appears meaningless. A bump here, another there. But as a whole, there is incredible efficiency to their rambles. How does the sum (the colony) form something greater than the collection of its parts (the ants)? Gordon explores questions like these and more.

Day to day in the desert, Gordon's work requires her to wake up at 4:30 AM and roam past hundreds of nests she's come to know over the years. Every few yards there is another. Ants may live for only a year or so, but colonies last twenty to thirty years. Females are the builders of the nest, foragers of food, defenders of the colony, and nurses of the young. They do not receive commands but rather take note of the ant beside them to determine what they should do next. Ants interact by touch and smell, bumping into each other and "sniffing" with their antennas, which are bent at the midpoint and wave around like arms. The fragrant subtleties of her sisters tells an ant who belongs to the nest and if they've been working inside or outside. Her actions also depend on the frequency with which she encounters these odors.

Patrollers, for example, are usually the first to leave the nest in the morning, while the foragers wait for their cue to leave. When no dangers are present, patrollers enter the nest and touch antennas with foragers who, after several encounters no more than ten seconds apart, depart in search of food. If a returning ant carrying food bumps into another forager near the nest's entrance, the outgoing forager will venture out; in this way more and more ants leave the nest in search of the proverbial picnic. A wave of activity glides through the community. These interactions are used "to activate" a behavior, says Gordon, which in this case is foraging. The result is a dynamic network of connections that coordinate the colony and sustain the queen. The interactions on a local scale (peer to peer) have consequences for the sisterhood as a whole.

"Nothing ants do makes sense except in the context of the colony," writes Gordon.[36] As ants search for food, they leave pheromone trails that dissipate over time and distance. The pheromone intensity is highest where a greater number of ants have passed over a trail or where many have recently traveled. If a forager has found a short path to food, ants will return and venture out more frequently on that trail, reinforcing the path more times than foragers on longer journeys, thus attracting more nest mates to the shorter path in a positive feedback loop. Over time, the path becomes the colony's prime highway. In this way, the most efficient route is determined not by one ant but by thousands of ant interactions. Many of us have seen this prowess firsthand in our kitchens—a line of ants marching into cupboards to forage for crumbs.

In a computer simulation of the ants' behaviors, Marco Dorigo, inventor of the Ant Colony Optimization metaheuristic and codirector of the artificial intelligence laboratory of the Université libre de Bruxelles, even showed that the colony's simple rules can be used to find an optimal solution to the devilishly challenging traveling-salesman problem, where finding the best path between cities amid many possible routes can seem like a Herculean task. The problem becomes considerably more tricky the more cities we add. If you want to earn yourself another brain wrinkle, consider this: a traveling-salesman problem with just ten cities has 362,880 possible routes, far too many for a computer to reasonably handle in time for schedule-crunched companies. Due to the problem's difficulty, scientists look to heuristics, or approximate methods, to speed up the time a computer takes to find an optimal solution. To do this, Dorigo programmed his simulated ants to travel a map and then exploited the colony's positive feedback loop, whereby shorter routes are strengthened by a greater number of single ant units.

These days, collective intelligence is a hot topic. There's even a journal called *Swarm Intelligence* dedicated to the subject. Flocks of birds, schools of fish, herds of wildebeests, pods of dolphins, colonies of ants, and swarms of bees have captured the attention of biologists, ecologists, mathematicians, engineers, and roboticists across the globe. Which brings us back to Southwest Airlines and its bottleneck problem, described at the beginning of this chapter: Can we prevent traffic jams when planes arrive at their flight gates? Doug Lawson, a Southwest systems engineer, turned to ants for inspiration. He designed a computer simulation of large-scale airline operations with "agents" that acted like ants. These ants (or each aircraft) learned over time from their virtual simulations which routes were fastest to use. Rather than pheromones, each aircraft remembered the best gate to arrive at and forgot those that took longer. Many simulations later, each plane had learned which gates to use for the quickest arrival. Every evening, the next day's activities were fed into the simulation and the strongest paths were singled out. This simple design lessened the time "antsy" passengers had to wait in order to deplane.

But Southwest didn't stop there. Pleased with Lawson's results, the airline asked if he could also apply swarm intelligence to their cargo routing and handling systems. According to the *Harvard Business Review*, planes were sometimes using only 7 percent of their cargo space, and yet at some airports, there wasn't enough space for scheduled freight loads. This led to bottlenecks in the airline's cargo routing and handling system, even when employees loaded freight into the first plane going in the right direction. Tired workers spent too much time moving cargo around. When there wasn't enough space for the loads, the backup rippled throughout their cargo routing system. It was messy and inefficient.

Lawson's algorithm found that sometimes it actually makes sense to leave cargo on a plane that is initially flying in the wrong direction. As Eric Bonabeau and Christopher Meyer explained in *HBR*: "If, for example, they wanted to send a package from Chicago to Boston, it might actually be more efficient to leave it on a plane heading for Atlanta and then Boston than to take it off and put it on the next flight to Boston."[37] This slight alteration cut transfer rates by as much as 80 percent and decreased workload by 20 percent. The improvement provided $10 million in annual gains.

These scenarios are inspired by just a few species of ant, but there are more than 12,000 species, found in almost every conceivable environment on Earth. Another striking illustration of swarm intelligence can be found in the ladders, chains, and rafts that fire ants construct with their bodies in response to a flood. They swarm and link their legs and mouths together to form rafts that can sail for weeks. A single industrious fire ant swept away by high rain will drown, but if they behave like molecules, merging to form a whole that is greater than its parts, they can survive. Unlike molecules, the fire ants shove their larvae underneath the raft to maximize the colony's buoyancy. The queen is placed at the center of the raft. Once they hit land, the fire ants immediately disband and get the queen to safety. It's an incredibly "intelligent" act.

Watching ants may sound tedious, but it is deceptively hypnotizing. While in the Mojave Desert, I tasked myself with watching their operations for an hour. Ants trickled in and out of their nest, which looked as if someone had poked a hole in the sand and created a mini-volcano of soil. One ant in particular caught my attention. She carried in her jaws what appeared to be a small dead ant. I settled in and decided to keep my eye on her. She hurried away from the nest, navigating rocks the size of her body with hardly a problem. On and on she went, past her peers; on and on I watched. She stumbled at a brown weed growing from a rough crack in the dirt, but she never let go of her dead sister. After about half an hour of this, the sun began to dim and I contemplated leaving her to her ways. Then she dropped her cargo, and as she scuttled away the "dead" ant popped up and plodded in the opposite direction.

What was this? I returned to the nest in search of another ant to follow and noticed the same thing happening: a bigger ant carrying a smaller one from the hole, still alive (despite appearances to the contrary). And again another—this one dropping her sister closer to the nest. Were these carried ants newborns? A different species? I wasn't sure, nor did I have a way to find out right then. I had entered a kind of trance. Just hours before, I'd read of a mob of protesters storming the United States Congress. They had broken through police lines, smashed windows, and stolen podiums; at one point they'd occupied the floor of the Senate, many armed with weapons. Lawmakers were told to wear gas masks and hide under their seats. Democracy was under assault. And here I was, in the middle of the desert, watching ants. It was surreal, the way psychedelics or deep meditation can feel. It was German philosopher Friedrich Nietzsche who wrote that "madness is something rare in individuals—but in groups, parties, peoples, and ages it is the rule."[38] That is the case sometimes, but we too can swarm on a good idea. Wikipedia, for example, is an accumulation of knowledge from a large group of people, and it manages to present fairly accurate information. Experiments also suggest that our language is possibly the result of swarm intelligence, our cultural self-organization responsible for its complexity and expressiveness.

Others who have used swarm intelligence include Unilever, for efficiency in equipment use at one of its production facilities. Telephone companies have tested it out to speed up calls. France Télécom, British Telecom, Hewlett-Packard, and MCI WorldCom have researched the use of ant-based routing methods. Owen Holland, a computer science and cognitive robotics professor in the U.K., invented a routing method that uses "virtual pheromones" at network nodes to reinforce uncongested phone call paths. Air Liquide used an ant-inspired method to organize truck routes delivering industrial and medical gases. The seemingly simple abilities on display in the insect world are also inspiring applications from online searches to groups of robots—exchanging information between individual units—to explore a planet or a toxic building. The added benefit of this last example? If one robot goes down, the entire mission doesn't come to a grinding halt.

If we think back to 1994—when an eight-legged, 1,700-pound robot named Dante II was teetering on the rim of Mount Spurr, an active volcano in Alaska—such a ruinous incident almost happened. (Dante II is so named because of another near calamity with Dante I, whose mission failed when a communications cable between the robot and its control station snapped.) The NASA robot, named after the poet who descended into hell in his "Divine Comedy," trekked over a hundred feet down into the volcano to sample gases spewed from the bowels of our planet. By the fifth day, the robot's laser-scanner mirror was filthy with volcanic ash, a precarious predicament in already difficult conditions that were further complicated by snowmelt. The robot needed visual images to assess its terrain and move its mechanical legs. As Dante II attempted to climb out of the crater, it toppled on its side halfway up, completely helpless. A tether on the robot was attached to a helicopter to lift it out, but the tether broke and Dante tumbled closer to the molten inferno. Ultimately, a human had to hike down the treacherous slope and attach a sling to Dante so the robot could be pulled out by helicopter. While Dante's mission was to gather gaseous data, the principal objective was to develop technologies that could lead to explorations of rugged planets in outer space. If such an incident had happened on another planet where rescue was impossible, Dante II would have become an expensive piece of litter. If a swarm of small robots had been sent to the crater, perhaps the mission could have been salvaged without human intervention.

Scientists like Australian roboticist Rodney Brooks think robots will continue to get smaller. Much of Brooks's own work, including the Roomba vacuum cleaner, is inspired by the simplicity of insects. With relatively few neurons compared with our own, bugs can find food, mate, and dodge a hand trying to slap them away. These bugs don't assess every situation and make a plan; they are driven by feedback loops that have evolved over millions of years. Imagine, Brooks says in the *Journal of the British Interplanetary Society*, a herd of micro-rovers for planetary exploration. "Upon landing either together or in smaller groups, the rovers would disperse covering wide ranges over the surface. Not all of the small rovers need be alike. Different ones could be specialized to particular scientific goals by the selection of (small) instruments onboard."[39]

Just like ants, the robotic mission would be decentralized and resilient. This stands in stark contrast to today's robots that are designed to work alone, not as part of a team. And yet the design of many micro-bots is simpler than one complex robot because the processors can be small and inexpensive. By applying the phenomenon of natural swarms to robotics, we can create flexible, robust systems that show cooperation and coordination. As Radhika Nagpal, a Harvard computer scientist who mimics collective intelligence in her robotic systems, notes, "the beauty of biological systems is that they are elegantly simple—and yet, in large numbers, accomplish the seemingly impossible. At some level you no longer even see the individuals; you just see the collective as an entity to itself."[40]

Some journalists have suggested that another application for swarm intelligence is in the treatment of cancer. This idea is likely overreaching. When I talked with cancer researchers, they dismissed the association, primarily because cancer cells are clones—a cancerous cell that doggedly duplicates itself into a dangerous tumor and beyond. There is no signaling, no interaction between players to coordinate movement and activities. As an alternative, a medical expert guided me to the immune system, where the analogy holds more power. The immune system has different types of cells (such as those in the blood, bone marrow, and lymph nodes) that efficiently work together without any central command. If we look at the immune system's B lymphocytes—a type of white blood cell—we find that each one has receptors that respond to certain pathogens. Some 10 billion of these B cells circulate in the blood at a given time, ready to sound the alarm if they detect an invader. When a patrolling sentry detects a pathogen (that is, their receptor matches an invader), it secretes tons of antibodies that can identify the invader. The antibodies flood through the bloodstream on a search-and-destroy mission.

The B cell divides at an increased rate, creating more daughter cells that also secrete antibodies against the invader. The daughter cells differ slightly from the mother cell in random ways due to mutations, and these daughters go on to create their own daughter cells based on how well they destroy the invader, and so on. As Melanie Mitchell, a professor of complexity and computer science, notes, this results in a kind of Darwinian natural selection process that improves the ability of B cells to eradicate invaders. However, there is no one cell in charge of guiding the process. Instead, a cell sparks a cascade of signals among cells that go on to produce a complex response.

And yet, parts of this process are still fuzzy. How exactly do cells work together to "learn" what threats are present? How do they avoid attacking the body? What happens when this system goes awry? Unlike cells, ants are observable with the human eye, which allows scientists to watch and learn how they act as a superorganism. As science writer Ed Yong notes in *Wired* magazine, "science, in general, is a lot better at breaking complex things into tiny parts than it is at figuring out how tiny parts turn into complex things."[41]

Yet this is exactly what this collective behavior reveals: how a hive mind can "think" without any guidance from a master.

Hive Mind

IF ANTS OFFER us a glimpse of swarm intelligence as it pertains to gathering food, bees show us how it can be used in a different way: to come to a unanimous decision on where to live. Bees must choose a well-ventilated site that's protected from the heat, not too exposed, high enough off the ground, and shielded from pounding rain and high winds, among many other considerations. It's a life-or-death problem, and yet a swarm of pin-sized brains need only a couple of days to come to a decision.

The first person to decode the collective intelligence of bees was German zoologist Martin Lindauer in the 1950s. Born in a small village in the foothills of the Bavarian Alps and the second youngest of fifteen children in a farming family, Lindauer lived a frugal existence. Years later, he was drafted into Hitler's work service and put to work digging ditches in a bog called the Dachauer Moos, where he was bullied for his lack of enthusiasm for Hitler's regime. Eventually, he joined the anti-tank unit of the German army. His combat experience came to an abrupt end when he received shrapnel wounds to his arm from an exploding grenade on the Russian front. During his recovery, his doctor recommended he attend a lecture by Professor Karl von Frisch on cell division. According to Thomas Seeley, a biology professor at Cornell University, Lindauer recalled two ideas rising like smoke from von Frisch's talk: "First, a new world of humanity, one in which humans create rather than destroy, and second, an encounter with science, an endeavor where humans use truth rather than lies."[42]

Lindauer started his PhD investigating honeybees with von Frisch as his adviser. His work involved diligent days of bee-watching, note-taking, and painting small dots on bees to mark their moves. His fascination with bee decision-making in particular began when he stumbled upon a small swarm that had buzzed away from their hive at the Zoological Institute in Würzburg. None of the bees had the customary lemon-yellow freckles of pollen sprinkled on their hairy bodies. Instead, they were behaving oddly, taking turns zealously shaking and wiggling their abdomens. Some were dusted with red brick flakes, others with gray soot from chimneys.

The bees were not lifting pollen from one flower and depositing it on the stigma of another. They were, he thought, communicating their explorations of nooks and crevices as potential new nest sites. He asked the institute's beekeeper to let him study the swarm. Her response was quick, and not what Lindauer wanted to hear: "That is out of the question, we need that colony."[43] After much negotiation, she reluctantly gave him permission. Bees choosing a new home is quite something to witness. Initially, the bees gather into a beard-like cluster on a tree branch, from which twenty to fifty scouts, usually the oldest bees in the colony, are sent out to hunt for a new nest. Each scout will then return to the group and waggle its body. The whole procedure takes on the appearance of a dance circle at a club, with new dancers entering the ring to show off their moves. In this circle, though, the bees are showing off not their "cool moves" but the potential new home they've found. With no vocal cords to rely on, the bees have found another way to say "over here." A scout bee moves into an open space and scuttles forward, jiggling her abdomen. The length and angle of the dance tell her peers the general distance and direction of the site she's found. When other bees exit the hive, they note the position of the sun and fly away from it at the same angle at which the scout danced.

If a scout bee thinks her peer may have found a superior site, she'll buzz off to check it out herself. If she agrees, she will return and join her comrade in a waggle dance, eventually recruiting others to the site. An uncommitted scout may visit many bees' sites before settling on one she finds worthwhile. It is as if the bees are holding a plebiscite on the colony's future home. Over the next day or two, the scouts settle on the best spot and rouse the rest of the swarm, steering them to their new domicile. Lindauer, ever the scientist, would sprint beneath the swarm as best he could to discover the colony's destination. Almost always, the bees come to a consensus that provides them with the best spot for a home among competing sites. Reflecting on his life's work later, Lindauer said, "I have to thank the swarm bees for the most beautiful experience in all of my projects."[44]

Lindauer's observations were confirmed years later when Seeley devised a simple but ingenious experiment on Appledore Island, off the coast of Maine. Bees love nesting in trees, and Appledore has almost none. The winged insects were ferried there for a special experiment where empty boxes were scattered across the island. The absence of trees ensured the bees would choose Seeley's boxes as nest sites, allowing him to observe how they selected a home among multiple choices. Seeley placed five plywood boxes on the island, four of mediocre quality that were slightly cramped, and one of excellent quality with plenty of space to thrive. Scouts would spend nearly an hour inspecting a potential nest site, buzzing near the entrance, assessing the size of the hole, and hovering all around the site to check it out. When inside, scouts scrutinized every corner of the hollow. Seeley's experiments suggest bees prefer a nest with an entry that isn't too large (so the hive isn't vulnerable to predators) but also not so small that bees can't fly in and out. They also prefer a bottom entrance to a hole on top, likely to prevent rain from drowning the nest, as well as a roomy hollow for the entire colony. It's an added benefit if the site is already furnished with some beeswax combs from a previous colony. When a scout finds a so-so nest site, she performs a waggle dance of only a dozen or so circuits. But if a first-rate site is discovered, she dances a hundred or more circuits, increasing the likelihood that others will see her dance and then check out the site themselves. The more she waggles, the more scouts are interested, and the more rapid the consensus for the best site. When the population dynamics of the bees were mathematically modeled, the best site was almost always chosen. Such behavior is nature's way of accumulating diverse perspectives and using collective wisdom as one hive mind. In this way, "the whole is more than the sum of its parts," as Aristotle said, and nowhere is this more true than for eusocial ("truly social") insects—those that live in a colony with only some individuals able to reproduce.

But what happens if there is a split decision in the colony? What if half of the scouts say, "I like this nest best," and the others say, "No, this one is definitely better"? There is no fighting. The bees resettle at their original location and continue their deliberations until one site takes the crown. When a scout is recruited, she does not simply trust her waggling peers but instead examines the potential new home for herself. This independence ensures the scouts don't propagate errors. Once the colony has come to a final decision, the cluster of waiting bees fly off to their future home and sculpt their glorious golden combs. "A swarm takes full advantage of its inherently collective nature to assemble rather quickly—often in just a few hours—a profusion of alternatives from which to choose," write Seeley and colleagues in *American Scientist*. "The larger this set, the more likely it includes a first-rate site. Thus, we see that one key feature of a swarm's decision-making is its decentralized organization, which helps ensure that it has a broad set of options."[45]

Such creatures have forced us to rethink our definitions of intelligence, or, at the very least, types of intelligence. Even slime mold, of all things, makes us reconsider collective intelligence in relation to the individual. So do fish. And birds. And locusts. A slime cell or a single bee in isolation isn't so smart. But swarm intelligence *is* smart. Some scientists wonder if the human brain works similarly; each neuron isn't so "smart," but the brain as a whole is. Is the brain itself a hive mind?

Researchers are even exploring whether swarm intelligence might be used to help AI cars navigate highways in the future. While bee and ant swarms have inspired real-world applications, for the most part, scientists are wary of using the collective behavior of ants or bees as some kind of guru life advice. There may be lessons to learn in their decision-making—for example, crowds are wisest if members act independently and responsibly without pressure to succumb to the wishes of others—but most scientists are more interested in understanding what their behavior can contribute to disciplines like medicine, robotics, and optimization technology.

Mark Kerbel, one of the founders of Regen Energy (now Encycle), came up with the idea for the company after experiencing the infamous 2003 blackout throughout the Northeastern and Midwestern United States. In the United

States, the buildings sector accounts for 40 percent of primary energy use and 76 percent of electricity use. The HVAC (heating, ventilation and air conditioning) systems account for around 35 percent of total building energy consumption. Kerbel, with cofounder Roman Kulyk, developed an algorithm that could cut unnecessary energy consumption for midsized buildings.

Typically, air conditioners, furnaces, and other appliances are ignorant of what is going on with other processes. Inspired by the communication of bees, Kerbel and Kulyk developed the EnviroGrid Controller, which sits on each HVAC unit's control box. Every couple of minutes, the controllers ping each other to learn the power cycles of each appliance. There are usually ten to forty controllers, or nodes, that work together within a given building. They then use Zigbee, a wireless technology, to link the smart devices to a network. In this way, the network can figure out the most efficient times to turn the systems off and on. A custom algorithm spreads out the energy demand and considers the situation of the other nodes in order to make the most energy-efficient decision. For example, if a refrigerator needs to turn on to remain at a minimum temperature, a controller connected to a fan or pump will shut off for a while in order to keep the building's total power output below a certain threshold. The devices work to satisfy not only the local constraint (i.e., the refrigerator) but also the system objective.

These are only some of the many ways Earth's so-called creepy-crawlies have inspired our technology. And yet, even after all of this research we still have questions: How do bees plan ahead, if they even do? When Swiss entomologist

François Huber and his team placed a glass barrier between where bees had started building their honeycomb and their intended end point, the bees curved their work to attach the comb to the nearest wooden frame (glass isn't a good surface for attaching wax). How did the bees know to do this *before* they reached the glass? Their behavior suggests a flexibility we do not fully understand. It's astonishing how much we are still learning, especially considering humans invented beekeeping around 10,000 years ago. Our sweet tooth for honey is depicted on Mesolithic walls, with paintings of humans collecting honey from wild bees. Chemical traces of beeswax have also been found on pottery vessels in North Africa dated to around 9,000 years ago. Then there is ancient Egyptian iconography of honeybees that adorns 4,400-year-old walls, and jars of the liquid gold found in the tombs of pharaohs such as Tutankhamun.

And yet as much as we treasure bees, we also ruin their populations. Commercial honey bees pollinate $15 billion worth of food crops each year, such as apples, almonds, and blueberries, and still beekeepers in the U.S. lost more than 45 percent of their honeybee colonies from April 2020 to April 2021, according to the fifteenth nationwide survey by the nonprofit Bee Informed Partnership. Such losses are remarkable considering wild bee colonies are sustaining relatively stable numbers. Some scientists pin the blame on industrial beekeeping, where bees are housed less than a few feet apart, rather than hundreds of feet apart as seen in the wild, to boost production efficiency. Supersized colonies nearly as large as a person is tall are also seen in commercial beekeeping, making these hives ideal hosts for pathogens and parasites. Varroa mites are particularly destructive, rapidly spreading from colony to colony and decimating hives. Yet these honeybees are critical to our agricultural success, with biologists endearingly likening them to "flying penises." Now, at least, bees can lay claim to a more austere epithet: "cyber inspirations."

Incredibly, swarm intelligence, while fascinating and practical, isn't the only thing the ant and bee world has to offer, for those who are willing to look.

To the Bug World and Beyond

CLINT PENICK IS a former punk musician turned ant scientist who is hoping the tiny critters can inspire us in the medical world too. His obsession with ants has deep roots: his first music band was called Army Ants (later changed to Kids Like Us), and he helped launch ants into outer space as a postdoctoral researcher at North Carolina State University. The ants were taken from a small mining town in North Carolina called Spruce Pine and drafted into the space program in 2014 as among some of our smallest astronauts yet. It was all part of a citizen science project for children to compare ants aboard the International Space Station with those in their classroom and document whether the astro-ants behaved any differently than the Earth-based ants. In other experiments, Penick has estimated that ants on Broadway in New York City eat the equivalent of 60,000 hot dogs per year. It's safe to say ants have been an obsession for him as long as his guitar, except his research papers get more coverage than his music. That's fine by him, he says. He does, however, miss the stage dives.

These days, Penick is an assistant professor at Kennesaw State University and studies ants' antimicrobial properties.

Social insects live in dense, teeming colonies of thousands to millions with a high probability of disease transmission. Some species even pass food to each other through mouth-to-mouth feeding, an ideal pathway for bacteria to exploit and decimate nests. Most ants in a colony are related, so if one ant is susceptible to a pathogen it's likely the others are too. To top off an already dizzying list of why ants shouldn't exist, their home turf is, or at least should be, a mecca for microbes. Their nests are warm, moist, bountiful—the perfect environment for unwanted bacteria to thrive. And yet, ants defend against such nuisances with a suite of homemade antimicrobials.

In a high-risk environment, ants have gained the upper hand after millions of years of evolution, and scientists are looking to their pathogen defenses as a source of inspiration for vital antimicrobials. More than 2.8 million people in the United States each year are infected with antibiotic-resistant pathogens; 35,000 of those people die. Public health experts warn of a global crisis in treating infectious diseases. Ant research, says Penick, could help scientists develop new antibiotics in the fight against human diseases. To investigate the antimicrobial abilities of ants, Penick's team used a solvent to remove any substances on the ants' bodies. The resulting solution was then introduced to a bacterial slurry, observed, and compared with a control group of bacteria. Of the twenty ant species they tested, twelve had an antimicrobial agent on their exoskeletons. A small yellow insect called the thief ant (*Solenopsis molesta*, named for its habit of nesting close to other ant colonies and stealing their food) had the most powerful antibiotic effect of all of them. The compounds from the thief ant showed no bacterial growth at all. Just as riveting, the team discovered ant species that don't even need to use antimicrobials. They must make use of alternative ways to prevent infection. "Rather than producing compounds that directly kill pathogens, some ant species might produce compounds or have physical structures that promote the growth of beneficial microbes," speculates Penick in his research article.[46]

One thing is certain: thousands of ant species exist, and they don't all ward off bacteria the same way. Some ants "self-medicate" and eat toxins like hydrogen peroxide when infected with a fungal pathogen.[47] Wood ants double up on their protective measures, collecting conifer resin with antimicrobial properties and spraying formic acid from their venom gland to disinfect their nest. Others produce antimicrobials in their bodily secretions.

Microbiologist Massimiliano Marvasi of the University of Florence and his team found that the antimicrobial secret sauce of a fungal-farming ant species was a dash of "transformation," by which he means antimicrobials that subtly change in structure and combination over time. This makes it tougher for parasitic fungi to gain an advantage, even after millions of years. Some ants make several antimicrobials at once. Marvasi says mixing or regularly varying antibiotic cocktails may hold promise as a means to curb antimicrobial resistance. The substances have yet to be turned into drugs, but the results are fodder for future research.

Today, Deborah Gordon continues to observe ants in the desert, and Thomas Seeley still works with bees on Appledore Island. Mark Kerbel passed away in 2017 after a long battle with cancer. Martin Lindauer died in Munich in 2008, his contributions to the science of insect collectives a lasting legacy, even when his own homeland was falling apart. More discoveries are likely to be made, but we are losing species faster than we can discover them. It's as if we are burning a library without bothering to explore what the books contain. Scientists have yet to take full advantage of what Earth's insects offer us, from solar-inspired technology based on the light-capturing ability of butterfly wings, to the ventilation systems of termite mounds, to the rubbery protein known as resilin found in insect joints that allows them to jump long distances and flap their wings millions of times. The springy material stores and releases energy, allowing for quick energy use, like snapping a rubber band. Resilin is more resilient than rubber, meaning it can be stretched over and over again without losing its properties, and could be used in everything from bouncy balls to spinal implants.

Finally, though this is certainly not an exhaustive inventory of insect-inspired work, there are algorithms. The way animals move and work as groups has inspired all sorts of computer scientists. Two of the most common algorithms are Particle Swarm Optimization and Ant Colony Optimization; there's also Artificial Bee Colony, Cuckoo Search, Locust Swarm, and so on. Even Google's search engine determines a page's importance by using the collective intelligence of the web. As Ed Yong notes in *Wired*, "biologists are used to convergent evolution, like the streamlining of dolphins or echolocation in bats—animals from separate lineages have similar adaptations. But convergent evolution of algorithms?"[48] Is there some basic rule that underlies collective intelligence? Perhaps yes. Perhaps no. There could be more than a single equation. But we may never know if we burn down the library.

5

A LEGGY TURN OF EVENTS

Giraffes Inspire Lymphedema Compression Leggings

"Nature is trying very hard to make us succeed, but nature does not depend on us. We are not the only experiment."
R. BUCKMINSTER FULLER, engineer and architect

THE STORY BEGINS with a big bang. Not *the* Big Bang, mind you, but the bang of a fist hitting a table in frustration. Her damn leg! Swollen like a sausage. With a wince, she swings her bloated right leg onto a chair and presses her fingers into the fluid-filled skin. She moves her hand and then stares at the indentations her fingers have left behind, watching as they slowly fill back up with fluid.

Hertha Peterson was born in 1920 on a winter's morning in Everett, Massachusetts, almost three months to the day after women were granted the right to vote. She was the only child of Swedish parents, and after excelling in high school, she attended Katharine Gibbs Secretarial School, which was known for the thousands of executive secretaries it churned out. They were well trained in office skills, organization, and punctuality and were renowned for their hats, heels, and white gloves. After graduation she was hired as an administrative assistant to the president of First National Bank of Boston.

In 1943, Hertha married Frank Shaw, and so began their sixty-three-year love affair. Hertha's husband graduated from Tufts University with a degree in electrical engineering and served as an officer in the navy during World War II. It wasn't long after the war ended that Hertha's leg began to swell. She had contracted secondary lymphedema (*edema* from the Greek "to swell") following an infection in her leg after the birth of their third child. At the time, however, she didn't know that. Little was known about the disease in the 1950s, and she was only correctly diagnosed after thirty years of living with the condition.

In the 1980s, Hertha was finally prescribed extra-firm compression stockings of 40–50 mmHg (a unit of pressure)—one of the highest compression classes in existence. But even these were too elastic to keep her on her feet. Every few hours her leg would swell to the point where she was forced to elevate her foot. This drained the fluid back into her upper leg—a temporary fix, but one that at least brought some relief from the pain. Her husband watched with a pain of his own: the ache of seeing his wife suffer with no resolution in sight. For years, Frank tried to find a solution. Countless ideas were jotted down, tweaked and trialed, and ultimately tossed into a discard pile.

Hertha was not alone. More than 150 million people worldwide suffer from lymphedema. The lymphatic system is crucial to the healthy maintenance of a body. If all is well, protein-rich fluid circulates throughout the system, collecting bacteria, viruses, and waste products along the way. Those unwanted dregs are then carried from the lymph vessels to the lymph nodes, the six hundred or so bean-shaped organs scattered throughout the body. In these little knobs, the waste is filtered out by infection-fighting cells called lymphocytes. But for lymphedema sufferers, damage to the lymph nodes results in an excess of fluid that swells the limbs and causes severe pain. For Hertha, this meant constant pain and the need to elevate her bloated leg on whatever piece of furniture was nearby.

Over the years, the Shaws visited the top vascular surgeons in the country, to no avail. A doctor in New York suggested they go to Germany to try a new inelastic bandaging technique, but the trip was too pricey for them. As he contemplated their options that day, Frank decided to take a breather from the sterility of hospitals and white coats and, on a whim, visited a giraffe exhibit at a local zoo. Reveling in the complete change of atmosphere, he gazed at this stunning creature with its long, dreamy eyelashes, flaxen stained-glass coat, cloven hooves the size of dinner plates, and, of course, that legendary neck. But it was the giraffe's slender extremities that really caught Frank's eyes. He considered the difference between his wife's inflated leg and the rod-thin limbs of the giraffe. As flocks of buck-toothed kids and child carriages made their way past, Frank stood and stared, mesmerized.

As the tallest animal on Earth, giraffes must have to withstand the highest leg pressure of any creature, he thought. At this pressure, they should be crippled with painful, swollen edema in their legs. But they're not. So how does a two-toed giant with a barrel stomach propped on skinny legs keep blood from pooling in its legs? Could the answer somehow help his wife?

Frank began to research.

Signature Mistakes

ON THE OTHER side of the world from the Shaws, the crackle of leaves dissipates in the air as a giraffe curls its tongue around a bunch of acacia leaves and strips a branch clean. The giraffe lives in the savanna, where its snacks are protected by thorns the size of sewing needles and vultures carve the sky with switchback turns. To thrive in this landscape, giraffes have evolved into marvels of engineering. Over the years, archaeologists have scrabbled together an evolutionary ancestry for the Giraffidae group, the members of which now include only giraffes and okapi. This has been no easy task. The giraffe's ancestors experienced a variety of transformations through the epochs. Pick one at random and you'll find a burly body with broad horns that resemble ossified ears (Shiva's beast). Pick again and you'll get a three-horned animal the size of a deer with small, saber-like fangs (*Xenokeryx amidalae*, named in homage to *Star Wars*' Queen Amidala).

The giraffe's name has also gone through many transformations. One of the earliest was *zemer*, a word derived from a Hebrew root that means "to prune"—a reference to their gentle habit of pruning trees of their leaves. The ancient Greeks called them *kamelopardalis*, a word that joins *kamelos* (camel) with *pardalis* (leopard), because the creature's neck is long like a camel's and its coat is spotted like a leopard's. The modern name "giraffe" possibly comes from either the Arabic word *zarapha*, meaning "fast walker," or *zerafa*, for "charming." Modern descriptions of giraffes swing from pegging them as the odd-looking ungulates of the savannas—muscled creatures with a vestigial awkwardness that makes them appear as if they've not quite left their youth—to creatures that move through tall grasses with an operatic grandeur. The disagreement likely stems from the moment when giraffes splay their scrawny legs in a rather inelegant show to take a drink, resembling Bambi doing the splits. As giraffes sprouted toward the skies, bending down to drink water became a handicap. They are at their most vulnerable to lion attacks while sipping from watering holes, and yet this odd combination of lanky legs, wide girth, and stupendous neck has won out over other designs in the most daunting of arenas: evolution's survival of the fittest. Still, when it comes to gleaning inspiration from Earth's animals, it's important to question whether their adaptations are a wonder to behold or an inefficient design that's the product of an evolutionary constraint.

"We get used to the idea that evolution is good at producing beautiful, elegant animals that look as though they've been designed. We forget that sometimes they're not perfect and there are imperfections," evolutionary biologist Richard Dawkins tells an audience of veterinary students in an episode of *Inside Nature's Giants*.[49]

Dawkins stares down at a dead giraffe on a cold metal table. As he speaks, anatomists slice open the thick, bristly skin of the neck to reveal a powerful ligament and a hunk of muscle. But they are searching for something more subtle. Buried deep inside the neck is a lithe nerve called the recurrent laryngeal that runs from the brain to the larynx.

The nerve resembles a slender noodle. It travels all the way down the giraffe's neck, loops around one of the main arteries in the chest, and climbs all the way back up again—for a total length of 15 feet (4.5 meters). And yet, mysteriously, the brain and the voice box are just inches apart in the giraffe's body.

"No engineer would make a mistake like that," says Dawkins.

That same error is buried inside humans too, though our necks are shorter, so the mistake can be measured in inches and not feet. Why does such an obvious error exist? A living creature is a legacy of its history, its double-helix DNA carrying a legion of stories embedded in its genetic code, many of which are destined to remain unknown. But some stories in our DNA *are* knowable. Science tells us that errors in design usually have to do with our ancestors. So to understand why the laryngeal nerve error occurs, we have to go back in time—way, way back, past all land ancestors and into the ocean.

In the fishlike ancestors of modern mammals, the nerve makes no error. It goes straight from the brain to the voice box. This works because fish don't have necks. But as our ancestors evolved, their necks stretched and the nerve got stuck under the main artery in the chest. During the slow process of evolution, the nerve never made the leap over the aortic arch; it just grew and grew, along with the necks. Dinosaurs like the towering sauropods grew a 90-foot (27-meter) nerve! All because a little artery got in the way. Scientists have called the nerve's path "a monument of inefficiency."[50] It's an absurd detour, but it's also an important lesson: nature makes mistakes too. When we study nature's designs, we probe with a scientist's curiosity, and we find that not all of them make sense. Some of these designs are mistakes passed down through generations; others are beautiful solutions to a problem at hand. The goal of the scientist is to determine which is which.

Evolution doesn't create a *perfect* system; it creates a *better than* version of a given system to adapt to a specific environment. The genome of animals is insanely complicated, and it's only in the last seventy years that we have begun to understand the basics of molecular biology. DNA was discovered around the same time that the first color television sets went on sale, cigarette smoking was reported to cause lung cancer, and Elizabeth II was crowned queen of England. The human body comprises 30 trillion cells, most of which carry DNA (except mature red blood cells, which lack nuclei). The low error rate for DNA replication is a tremendous achievement on the order of one mutation for every 30 million base pairs (independent editing mechanisms check for errors). Try copying a book word for word without making a single mistake, even with a couple of people checking your work, and you'll get a sense of the monumental task at hand. DNA passed from one generation to the next usually accumulates just one hundred to two hundred new mutations. Despite this remarkable fidelity, changes and mistakes do happen. Scientists found that not only is our code extremely efficient at minimizing mutations, but when these errors do occur, they are unlikely to be instantly deadly, helping to up the pace of evolution. The errors that don't kill us live on and, if they are beneficial, are reproduced. All of us are continuing to adapt, even now; we never truly reach a state of perfection.

Perhaps the most recognized example of nature's mistakes is physical mutations. Take two giraffes born with dwarfism in 2015 and 2018; each grew to a height of 9 feet (2.7 meters), just a few feet taller than a typical giraffe's neck. It would be nearly impossible for a male of this height to mate with a typical female. Such mutations and evolutionary faux pas are a testament to the foibles and follies of life on Earth, without which we may never have had the fortune of experiencing the astounding diversity of creatures found far and wide on each and every continent. Examples of mutations that have become a blessing include almond trees, where a single genetic mutation thousands of years ago switched off the toxic compound amygdalin, which breaks down into cyanide when ingested, making it possible for humans to cultivate the tree to this day. There is also a small proportion of people who have a genetic mutation that gives them resistance to HIV, a virus that causes AIDS. And then there is a blood disease called sickle cell anemia that is arguably a "bad" mutation, except that people with it are more likely to survive malaria.

Giraffes have a complicated evolutionary history, too, one we still don't fully understand. Why do they have such long necks? We have ideas, but no absolute agreement. The leading theories are that males use their necks to fight for access to females, or to reach leaves that other herbivores can't reach. The argument *against* the fighting theory is that females have necks that are proportionally as long as males', and yet females don't fight with other giraffes. There are other questions, too. Why do giraffes have nubby horns on their head? The fighting theory crops up again here, but these little nubs don't grow large compared with other horned animals. Why are their tongues blue black, and not an ordinary bubblegum pink? To keep their tongues from being sunburned, one theory suggests, but we don't really know. Taken together, these theories reveal a startling fact about Earth's tallest creature: we don't know them very well at all. Which brings us back to our first question—the question that Frank Shaw asked as he stood at the zoo, staring at the giraffes: Why do their legs not pool with fluid?

Many years earlier, another man pondered a similar question about the improbable animal. August Krogh, a Danish physiologist who won the Nobel Prize in 1920 for his studies of capillary blood vessels, said, "It would be extremely interesting to know just how the giraffe avoids the development of filtration edema in its long legs." Giraffes withstand great pressure in the capillaries of their legs due to the high volume of blood weighing on them. This, in addition to gravity, should force fluid out of the capillaries—but somehow, it doesn't. "For a large number of problems there will be some animal of choice or a few such animals on which it can be most conveniently studied," said Krogh in the *American Journal of Physiology* in 1929.[51] This concept became known as Krogh's Principle.

The most famous example is the North Atlantic squid's axon—a thin fiber that branches from a neuron in the brain and transmits electrical signals. To "talk to" each other, neurons send electrical impulses down the length of their axons. The human brain is overwhelmingly complex, with just a piece of brain the size of a grain of sand containing 100,000 neurons and 2 million axons, all communicating with each other. Any given neuron may make as many as tens of thousands of synaptic connections with other neurons. Given these figures, the number of permutations of a brain that are theoretically possible is staggeringly high.

In the 1930s, when neurophysiologist John Zachary Young discovered that the North Atlantic squid's axon is nearly a thousand times wider than a human's axon, he quickly realized the squid was a scientific windfall, a godsend for those trying to study the minuscule complexity of the human brain. The squid's axon is about a millimeter in diameter, or roughly the width of a very thin capellini ("angel hair") noodle. The axon is so big scientists were able to thread an electrode inside and measure the axon's electrical current—a feat they could not do with a human's axon. In 1963, scientists Alan Hodgkin, Andrew Huxley, and Sir John Eccles won the Nobel Prize in Medicine for their discoveries on how electrical pulses fire in response to a stimulus, helping to unravel the nature of communication in the brain. Later, Hodgkin joked that the Nobel Prize should have gone to the squid, whose giant axon made all their work possible.

If we return to the giraffe, physicians during World War II, in pursuit of aviation medicine, wondered why the animals do not get head rushes or dizziness when they lift their heads rapidly. Every time the giraffe dips its head to drink, it risks blood rushing twenty feet—fast!—to its head, causing a stroke. When it swings its head up again, it risks fainting. If you were to suddenly rise after lying on the grass on a warm summer day, you'd likely feel dizzy too. And you'd only be lifting your head roughly six feet. Giraffes have the highest blood pressure of any animal, and as such they need to exert strict control over it. Not since the dinosaurs has a creature had to face such a formidable challenge.

It turns out that the animal manages this predicament with ingenious biomechanics. For years, scientists pinned the solution to these questions on the giraffe's twenty-five-pound heart. In order for the heart to pump enough blood around the giraffe's body, from the legs all the way to the brain, the blood must overcome significant forces of gravity. "Everybody believes that giraffes have a very big heart," says Christian Aalkjær, a professor at Aarhus University and Copenhagen University who explores the physiology of the cardiovascular system in humans and, more recently, giraffes. "But when we looked into the literature, there was basically no real measurement."[52]

Aalkjær's team found that in most mammals, the heart weighs half a percent of the body's weight; this is also true for giraffes. The giraffe's heart is comparable to their weight, no "extra-large size" after all. However, giraffe hearts do have incredibly thick, muscular walls that squeeze and relax, providing the pressure needed to move large volumes of blood. A series of valves in the jugular veins also help, restricting venous blood from exiting the brain too quickly and thereby maintaining the balance between the pressure inside and outside the brain's blood vessels. But it's in the giraffe's legs—those skinny, knobby things—where nature's prowess truly takes center stage.

"Why don't they develop edema?" It's Aalkjær's turn to ask this question, directing it at me the way a teacher would to a student. "If we had such a high pressure in the legs, we would develop edema," he says. "Why is it that the giraffe's legs are always elegant and thin, and not full of tissue water?"

To thrive at their size, giraffes have to withstand the highest leg vein pressure of any creature in the animal kingdom (250 mmHg). A giraffe's tight, inelastic skin acts as a permanent support stocking to prevent blood from pooling in its legs while also allowing fluid to briefly visit the interstitial spaces, or the spaces that surround tissue cells. Capillaries run through these tissues and, in order for nutrients to move in the body, fluid enters and leaves the interstitial space via the tiny blood vessels. It is at the boundary walls of this netlike system of capillaries that the exchange of nutrients, waste products, oxygen, and carbon dioxide takes place. In the adult human body, there are around 10 billion capillaries that, if placed end to end, could wrap around Earth twice. A giraffe has considerably more capillaries exchanging fluid throughout the body. Some of this fluid doesn't reenter the capillaries but instead enters the lymph vessels and is returned to the heart. Edema, as you may recall, occurs when this fluid in the interstitial space is not returned to the veins or the lymph vessels. This causes a person's lower legs to swell with fluid. In giraffes, however, this never happens—thanks to that tight, inelastic skin. As the giraffe gets older and bigger, its skin and vessel walls get even thicker.

Cutting into this skin "is like cutting into linoleum," says Aalkjær. As soon as the smallest amount of fluid leaks out of the blood vessels, it stops. There's no elasticity to accommodate the extra fluid. "It's as if they're born with compression stockings."

Frank Shaw discovered this kernel of knowledge during his scrutiny of the literature, and it sent him headlong to his next destination: his wife's closet. He took one of her best pairs of high boots and cut off the front panel. On one side of the boot, he attached bands of Velcro; on the other side, he fixed metal D rings. The Velcro looped through the D rings and doubled back on itself. He added five cinching bands in total: two on the calf, two on the ankle, and one on the boot's toe box. Each band touched another band to prevent the skin from bulging between the gaps. In this way, he was able to tighten the boot while using stiff material like leather so the leg couldn't expand. This closure method also meant the boot could be adjusted to the person, rather than needing exact contour measurements. A person could cinch the straps near the toes tighter than the straps near the calf to provide graded pressure.

"I won't wear them," Hertha reportedly told her husband, glaring at him in defiance when he first showed her his new idea.[53] He had ruined her best pair of boots, and for what? Nothing ever worked. They had seen the top doctors in the country. Why bother to keep trying?

Eventually, his pleas weakened her resolve. She brought the modified boots on their vacation. But she didn't wear them—not at first. Only after frequent rests to put her leg up and her husband's constant badgering did she finally succumb to the pressure and don the contraption. She walked around for the rest of the day. She didn't stop once for a rest, and she never winced in pain. For the first time in years, she felt an inkling of hope. Hertha turned to Frank and said, "Get a patent."

On November 5, 1974, Frank received patent 3,845,769 for what he called his Therapeutic Boot. Four years later, he launched Shaw Medical. With his daughter Sandra on board, the company began to research and develop lymphedema products. By 1990, Sandra had started CircAid Medical Products, Inc., and she purchased her father's inelastic, adjustable compression technology in 1993. Similar products have since entered the market, but Hertha continued to wear CircAid Leggings for the rest of her days.

Giraffes are now scarce across the land, their populations able to withstand the tests of time but not the cutting edge of a human blade. Poachers rarely hunt them for their meat, instead opting for something a little less caloric—their tails. Their tufted appendages are status symbols of authority, often used as regal flyswatters. It is a dismal sign of the times that humans may be their downfall, cracking off another branch on the once diverse Giraffidae family tree.

Giraffes have found a way to survive the harshest pressures of life, and as Frank Shaw discovered, we still have much to learn from these animals—lanky legs and all.

Lunar Spills

IF YOU SEARCH for Dava Newman on the internet, one of the first photos you'll come across is of a blond woman with an athletic build posing in a white skintight suit that's crisscrossed with veinlike lines. She wears a transparent helmet in the shape of a bubble and black kneepads, her matching black-gloved hands propped on her hips in a confident pose. It looks like a scene from a futuristic movie, but Newman is not an actor. She's an Apollo Program Professor of Astronautics at the Massachusetts Institute of Technology (MIT) and a former NASA Deputy Administrator, nominated by President Barack Obama. And she's modeling an astronaut spacesuit that she hopes will one day go to Mars.

Daughter of a pilot, Dr. Newman remembers watching the Apollo 11 moon landings in Helena, Montana, through the static of her television when she was five years old. From an early age, she was drawn to exploration. In 2002–2003, she sailed 36,000 nautical miles with her husband, Guillermo Trotti, and circumnavigated the globe. Now, standing in front of a TEDx audience on my computer screen, she showcases "bloopers" from the Apollo moon missions: astronaut Jack Schmitt of Apollo 17 in a marshmallow-white spacesuit fumbles with a sample collection bag on the moon and topples forward, a slow-motion wipeout onto his stomach and hands. Charlie Duke of Apollo 16 takes a leap in the lunar air, loses his balance from the weight of his suit, and falls on his back in what he describes as a near-death experience; had the fiberglass shell of his life-support system split open and damaged one of the pumps or other mechanisms, he would be dead. A tear in his suit and the only protection between him and the vacuum of space would be gone. The air inside supplies the pressure astronauts need to survive on the moon, but the trade-off is that they move like "stiff balloons," says NASA. It's particularly difficult to use their gas-filled gloves during repair missions. "Clumsiness" is practically sewn into the design of their suits.

Montages of these lunar spills are no secret and are easy to find online. NASA has reports for each puffy moonwalker who has tumbled, tripped, or fallen in the name of space exploration. A report of Commander David Scott's fall during Apollo 15 reads: "His right foot steps into a small depression and he begins to lose his balance. As he steps with his left foot, it slides off a small rock and continues sliding on the loose surface soil. While trying to drive his feet back under his center of gravity, Scott increases his forward velocity. He then falls forward with both hands extended to break the fall. Landing on his left side, he rolls counterclockwise and on his back and is then out of view of the TV camera."[54]

If astronauts were as clumsy on Earth as they are on the moon, they would never have made it up there in the first place. NASA reports four different falls on Apollo 16 related to astronauts picking up objects. Two additional falls happened when the astronauts inserted a penetrometer, a rodlike device that measures the force needed to penetrate the soil at various depths, into the ground. "Analysis indicates this requires maximum flexing of the suit...," reads the report. "This type of activity also requires sufficient experience and training to give the crewmen complete feel of the effort and coordination needed for the performance."[55]

The blunders described here are less a product of gravity than of the astronauts' lack of mobility. In effect, the spacesuit acts like a mini–home habitat, providing astronauts with the pressure, oxygen, and stable temperature they need to perform tasks outside the station. Once the astronauts have left Earth's atmosphere, there is no oxygen or atmospheric pressure. Everest is at 29,000 feet (8,800 meters), where the air becomes much more thin. Above 63,000 feet (19,200 meters), humans must wear specialized spacesuits to protect their body's fluids from vaporizing into gases. These suits are pressurized at 4.3 pounds per square inch (psi), which is considerably below the atmospheric pressure of sea level on Earth (14.7 psi), and even below that of Everest (4.8 psi).

The downside is that such spacesuits are clumsy and tiring. The gloves make fine motor movements tricky, and every object that is accidentally dropped in space is a danger—a projectile that could collide with a fellow astronaut or the space station itself. If an astronaut drops something, they need to report the object, its approximate velocity, and the direction of travel. Some dropped items include a spare glove in 1965, a thermal blanket in 1998, a spatula and a camera in 2006, pliers in 2007, a crew lock bag in 2008, a fabric shield in 2017, and a wire tie in 2018. The rogue spatula was not dinner gone wrong, but NASA's attempt to experiment with a thermal tile repair in orbit after a broken heat shield led the Space Shuttle Columbia to break apart on reentry. The spatula was used to spread goo into the tiles.

Newman hopes to improve astronaut mobility by redesigning their spacesuits. What's needed is a suit that provides 4.3 psi and more mobility than past moonwalker suits. The suit she can be seen modeling on the internet is, in part, inspired by the same marvel of design that led Frank Shaw to cut up his wife's best boots: the giraffe. As Newman writes: "Nature is often a powerful teacher."[56] After a decade's research into how to engineer a better spacesuit, Newman came up with the BioSuit. Her design's "second skin" is made of memory alloys with a web of thin, muscle-like wires and coils. To activate it, an astronaut plugs it into a power supply on the spacecraft to contract the coils and tighten the suit. The pressure produced is enough to support an astronaut in space, according to Newman. When the astronaut wants to squeeze out of the spacesuit, the coils are cooled to loosen them. It is, essentially, a controllable compression garment.

The BioSuit also features "lines of non-extension" (LONE), in which tension elements are positioned along lines on the body that do not bend or stretch. The concept was first developed in the forties by Arthur Iberall, an American physicist and engineer, for the U.S. Air Force and the National Advisory Committee for Aeronautics. Iberall pioneered a way to maintain pressure in an astronaut's suit after discovering that certain body points do not contract or stretch much despite a person's movement. When he connected the LONE points, he could apply the necessary mechanic pressure to the astronaut's body without compromising their motion. Test pilot Scott Crossfield flew in the suit in the X-15, a hypersonic rocket-powered aircraft, as did astronauts on some Gemini missions in the 1960s. However, there were issues with uneven pressure and the system was abandoned.

Newman has followed up on the design for her next-generation spacesuit, but this time she's working to provide uniform pressure over the entire body. The red, veinlike lines on the spacesuit are the LONE filaments that "shrink-wrap" the suit to the skin but wouldn't, for example, impair the movement of explorers on the surface of Mars. If our species hopes to one day establish base camps on the Red Planet and search for life, we can't have astronauts struggling to bend down or put a measuring tool in the ground.

Newman's research has led to another application as well. She's partnered with the Boston Children's Hospital and other scientific laboratories to see if the BioSuit technology can be adapted as a potential assistive device for children with motor impairment, such as those who have suffered strokes or those with cerebral palsy. In the future, Newman hopes to add actuators to the suit that could enhance and help direct movement.

More examples of such animal inspirations are on the horizon. Recent ungulate-inspired technology veers into the realm of camels. A cooling technology inspired by how the furry camel stays cool in the desert sun could be used to keep food or medical supplies cold for longer with less energy. Engineer Jeffrey Grossman of MIT is leading the experiment, which layers hydrogel and aerogel to mimic the function of the camel's pores and fur.

Scores of species lineages have experimented with designs to enhance survival, be it to stay warmer in polar regions or, like the giraffe, to stand tall in the savanna. We won't find answers to all of our questions by studying the various iterations of life on Earth. Their goals are not our own. Still, as Aristotle once wrote, "in all things of nature there is something of the marvelous."

6

BONDING WITH NATURE

Blue Mussels Inspire a Nontoxic Glue

"The world would suffer, today, much less in its comforts and conveniences of living from a loss of all its gold and silver than from that of its glue."
HORACE GREELEY, newspaper editor, 1872

ON A BLUSTERY winter's evening near the sea at Pillar Point in Half Moon Bay, California, a group of us squeeze into our rubber boots as a trickling of sandy, sunburned beachgoers trek back to their cars in sandals. The beachgoers had arrived earlier in the day when the tide was high and brimming with white foam. As they lounged on their towels, they saw boats sail on the horizon and seagulls fly like paper planes in the humid air. They probably didn't see, however, the many critters living near them that day, just feet away in the shallows. To see the coming and going of the tides is to witness the gravitational tug-of-war between the Earth, the moon, and, to a lesser extent, the sun, setting the stage for a unique ecosystem known as the intertidal zone. Here, creatures need the fortitude to withstand both the smashing of the waves and the baking-hot dry spells, all in a day's time.

Our group is here with the California Academy of Sciences to document what we find, our jackets flapping wildly in the breeze and our heads bent like diligent students to scour the rocks. Pockets of tide pools reflect the radiance of the setting sun like shattered gold mirrors. A man in a yellow rain jacket stares at his feet and then bends down on one knee. Water ripples in the gusts of wind as he peers past the surface to the creatures beneath. Everything seems to move and change in the intertidal zone: crabs scuttle sideways into crevices, sea stars raise a chubby arm from their quintet of tentacles, and shrieking seagulls scavenge the rocky flats. Even the "sea vomit" sways its slimy body to the tidal rhythm. This is the primordial divide, a place where the elements of earth, water, and air meet and facilitate change; where life can transition from sea to land as our distant ancestors once did. The intertidal zone calls to mind the Spanish saying "*Poco a poco se va lejos.*" Little by little, you go far.

I kneel and take a closer look at the only creatures that haven't moved: hundreds of mussels with twin blue-black shells moored to the rock. Every cranny is occupied by clusters of them, shells clamped shut and crammed side by side with their brethren. Blue mussels do move, says J. Herbert Waite, but it takes patience to witness. Waite is a professor at UC Santa Barbara who began studying mussel biochemistry as a bright blue-eyed graduate student in the 1970s and soon earned a reputation as a trailblazer in uncovering how mussels stick themselves so firmly to wet rock. When I met Waite in winter of 2020, there was something a smidge crotchety about this man who'd spent decades studying a creature that doesn't move much. But this first impression soon melted away, replaced by a sense of something more charming and idealistic.

In conversation, Waite's surly words and dreamer notions ebb and flow like the tides. He grew up on the seashore of Barnegat Bay in New Jersey, an intercoastal marine environment that doesn't have any big waves—perfect for rowboats and little motor skiffs. He spent his early years fishing, collecting, and lounging about in nature. His college years and beyond were about reconciling his love of biology and chemistry. He often said to himself, "To hell with *Homo sapiens*. I want to know how other creatures work as well." But he was practical enough to realize he "would have to make an argument of relevance to *Homo sapiens* to get funded."[57]

Like the creatures he studies, Waite's not into noisy fanfare; he's a quiet steward and scientist of the shores. He leaves the inventions hatched from his ideas to others. For him, curiosity and creation are engaged in a delicate tango. His philosophy is simple in principle: Basically, there are meaningful and not so meaningful ways to apply knowledge, particularly in the Western mindset, in which so much of invention has to do with the creation of convenience, not necessity. Eventually, he says, we get addicted to the convenience and it becomes a requisite. "It's fundamentally all baggage."

When Waite first dipped his toes into the world of mussel adhesives, nontoxic glues that excel underwater didn't exist. And yet, right in his own backyard, there were mussels that made a renewable, biodegradable glue in five minutes, a feat we struggle to replicate. At the heart of the mussel's success, says Waite, is "chemistry, the queen of the sciences." The story of chemistry is, in its own right, a scripture of the universe—a molecular blueprint for how all things are arranged and built.

If you happen to be watching a mussel at just the right time, says Waite, you'll notice it does some fancy footwork, a brown "foot" slithering out from its shell in search of a surface to call home. It's a strange, alien-like process to watch—this protracting and probing—and one that those of us not standing in a tidal pool can witness via high-speed video footage online. Upon striking a rock or boat hull, the mussel's foot makes stringy fibers called byssal threads that tether it to the surface like tent lines (cooks call these threads "mussel beards"). Dozens of these lines are forged one by one, and together they work to prevent the mussel from getting swept out to sea. These threads can stretch up to 160 percent of their original length and remain five times stronger than our own Achilles tendon. Juveniles use the threads like climbing ropes to haul themselves toward a better home. Once the thread is stuck, the foot retracts back into the shell.

The mussel's threads must survive the scorching heat of the sun and the erosion of seawater. It's an inspirational feat of endurance. As any surfer knows, it's a harsh world out there, a life of two existences: terra and sea. And yet, a niggling doubt crept into my mind as I watched the mussel: How much is there, really, to learn about a boneless, brainless bivalve? Does their glue differ substantially from our own? Do we even need a new glue?

Glued to Glue

MOST OF US never really stop to think about life before glue. Even more strange is just how gory and grim the tale of glue is. When we look around, it's easy to see how addicted we are to the stuff. We love to stick things together with glue. We love to build things with glue. We love to seal things with glue. We have glues for packaging food cartons, sealing envelopes, adhering aircraft pieces together, sticking soles to shoes, and making furniture. As I look around, almost everything near me is glued together—my pencil, my notebook, my cell phone, my hair tie, and even the sticker on my Honeycrisp apple.

Since ancient times, humans have harbored a desire to stick things together. Two hundred thousand years ago, Neanderthals burned the bark of birch trees over a fire to make a dark, tacky substance that they used to fasten handles to tools and weapons. Their creativity flourished more than 100,000 years before *Homo sapiens* in Africa ever thought to use tree resin and ocher adhesives—a revelation that certainly unsticks the Neanderthals' reputation as oafish, unthinking beings. One of the best-known examples of preserved birch glue was found with Ötzi, a frozen mummy who lived roughly 5,200 years ago and was discovered facedown by hikers in the Ötztal Alps near the border between Italy and Austria in 1991. Upon further inspection, scientists found that three feather halves were glued to the ends of his arrows with birch tar.

Fast-forward thousands of years to the beginning of recorded history and animal blood is prized by humans for its ability to dry into a strong, relatively water-resistant glue. Horses and cattle past their prime were said to be sent to the "glue factory," where the collagen in their hooves, skin, and bones could be transformed into adhesives for the paper and furniture industries. The animals were washed to remove dirt and then boiled to convert the collagen into a glue solution, which was concentrated by evaporation. In the early days, these factories were not pleasant places to work. In 1925, chemist Thomas Lambert wrote that where the bone factory is located is a matter of great importance. "In choosing a site for the erection of such a works, a position outside the boundaries of a town should be decided upon, in order that the offensive smell which arises from a works of this character may not give cause for complaint from a populous community."[58]

Collagen (from the Greek *kolla*, meaning "glue," and *gen*, denoting "producing") is indeed a useful ingredient: it can be turned into a gelatin that sticks when wet but hardens when dry. The candy industry is obsessed with it, breaking down pigskin, cattle bones, and cattle hide to glean collagen for gelatin delights they shape into little bears, soda bottles, and "fruit" snacks. Historically, even sea creatures were not spared from the hunt for collagen, their skins and bladders used in bonding rubber gaskets and paper to steel. New glues enter the marketplace all the time, including "meat glue" to stick meat to more meat. Once considered a tool for food manufacturers to use in the making of reconstituted products like chicken nuggets, deli meats, and imitation crab, meat glue has been whipped into fashion by haute cuisine chefs who create edible multispecies art (think bacon-cod medallions). Glue has been a quiet player in the background of modernity's rise, its macabre origins hidden by the sheen of its accomplishments. The popularity of blood adhesives rose fast between 1910 and 1925, as they were used in plywood for military aircraft, but by 1960 that market hit its peak at 50 million pounds per year, after which it steadily waned to near zero. Formaldehyde glues took the throne and became emperor.

Formaldehyde is a colorless, strong-smelling gas used as a preservative in mortuaries and as glue for pressed wood products such as furniture, cabinets, and paneling. It's a useful substance derived from petroleum and natural gas, but also a potential human carcinogen. Formaldehyde vapor is released in the production of plywood and particleboard and can irritate the eyes, skin, nose, and throat, raising toxicity and cancer concerns for both woodworkers and consumers. Despite formaldehyde's drawbacks, its contributions have been considerable, and the sheer volume of it can't be denied: in 1998 (the last recorded date), 11.3 billion pounds of formaldehyde were produced in the United States.[59] In North America, more than 90 percent of the adhesive resin used by the wood-products industry is formaldehyde based, and it is often created using fossil fuels. The business is booming: the United States and Canada spent more than $7.4 billion just on wood adhesives in 1999. Even in the RV where I lived out in the desert it's inescapable. The vehicle is a moving box of the stuff: the floor, paneling, cabinets, furniture, and countertops all have formaldehyde glue. Three vibrant red stickers inside warn of possible exposure.

Glue seems to hold modern societies together, which makes "stickiness" a big business. But some people are not so happy with our glue decisions, feeling we can do better to make a less toxic glue for both the planet and people.

Kaichang Li, an associate professor at Oregon State University with postdoctoral work in nontoxic composites, is one such person. While scouring blue mussels on the Oregon coast with friends one day in 1999, Li became captivated by how the creatures resist the tug of tides. While his friends were scrambling the craggy outcroppings and chucking mussels into buckets, Li, ever the scientist, bagged some for future research. Back at the laboratory, he set out to examine their super glue, sifting through the work of those, like Waite, who'd come before him. He quickly learned that we already know a good deal about blue mussels and the chemistry of their glue. But the work wasn't finished. Harvesting mussel proteins isn't feasible from a cost perspective. It would take 10,000 mussels to create one gram of adhesive proteins. Since we can't extract glue from mussels like we can milk from cows (nor, arguably, would we want to), Li needed to invent a synthetic version that mimics the bivalve's tenacious grip.

Wet, Rough, and Dirty

MUSSELS LIKE TO stick themselves where it is wet, rough, and dirty. Our glues perform best on dry, smooth, and clean surfaces. One of the great challenges for adhesive scientists is to create a glue that can withstand the presence of water, salts, and contaminants while staying sticky. Potential uses for such a glue range from surgical wound closure in a salty, wet environment, to repairing fetal membranes, to boat fabrication, and possibly even to helping repair coral reefs using a bio-friendly glue to transplant healthy coral back into the wild. Where we have come up short concocting these glues, marine mussels have mastered the art of sticking to a dirty, wet surface. For a mussel stuck on a rock, the surf's velocity in combination with the rubble in the water creates the effect of being sandblasted by debris. Evidence of this harsh treatment can be seen in the mussels' shells, which are often stripped clean of organic matter. The creatures thicken their shells in response to this constant battering—an adaptation that protects their bodies but not their threads, which poke out nearly two inches beyond their shell's protection.

How do the mussels manage to preserve their waterproof glue, both while it's being made and after it's stuck? The answer lies in the clever design of the mussel's wormy foot. Inside is a hollow tunnel, and the tip of the foot acts like a plumber's plunger; when the mussel puts its foot on a rock, it creates an airtight and watertight vacuum chamber. Not only does this secure a space devoid of seawater, but it also creates a force that helps draw liquid proteins into the chamber. The mussel then pumps and massages this foamlike substance into a threadlike shape. The process is similar to injection molding, where molten polymer is squeezed into a mold and given time to mature; the mold is then split open and, voilà, you have a sufficiently cured object. The rest of the curing comes with exposure, which in the mussel's case is provided by the seawater.

"Some of the curing processes are triggered by salinity but most of them are triggered by the difference in pH," says Waite. "Seawater has a pH of about 8," whereas the mussel's stockpiled proteins in the cell are usually stored at about pH 5. "So the difference between the two is a really powerful trigger for hardening." The mussel repeats this activity again and again, one thread at a time. "It's a very efficient little applicator machine in that sense."

That's the how of it, then. But what about the what? What mix of chemicals is getting sucked down the foot's chamber? Proteins in the mussel's threads are generally called mussel adhesive proteins, or MAPs, and each of these proteins contains amino acids. One amino acid in particular has received attention for contributing to the mussel's stickiness (especially in combination with other MAPs). Called L-DOPA, the amino acid is a close relative and precursor of the neurotransmitter dopamine, a chemical well known for its role in influencing mood and pleasure. When L-DOPA was discovered in mussels, it was primarily known as a critical therapeutic for Parkinson's disease, a brain disorder that causes shaking, stiffness, and difficulty with walking. Parkinson's is caused by a scarcity of dopamine, which allowed scientists to develop drugs to lessen the symptoms of the disease, though the disorder remains incurable to this day. L-DOPA famously became known as the drug in Oliver Sacks's *Awakenings* stories, where patients regained consciousness after decades of "sleeping sickness." Each patient responded differently to the drug, and some, like Rose R, awoke to the horrifying realization they were now thirty years in the future.

But in the mussel's case, "the L-DOPA is actually introduced into a protein, and that really makes it very alien to anything in Parkinson's disease," says Waite. In the mussel's foot, L-DOPA forms extensive cross-links and helps with stickiness. In particular, the synergy of L-DOPA and another amino acid called lysine is what makes the mussel's threads super strong. If you enlarge the chemical players to life-size representations, L-DOPA is the sticky Spider-Man and lysine is the trusty sidekick. Lysine prepares the wet surface for L-DOPA, like a primer before gluing. Lysine's positive charge kicks away positively charged ions on the rock, like magnets whose similar poles repel each other. With the all-clear, L-DOPA deploys its sticky powers to the surface—a perfect one-two punch.

In the laboratory, Li confirmed the sticking power of L-DOPA with wood, but his scientific explorations were far from over. He now needed to recreate the mussel's glue in the lab. Of course, the adhesive had to be cost-effective, but he also had another complication on his hands: the mussel's liquid proteins form a solid at room temperature, which doesn't bode well when trying to squeeze glue out of a tube.

If Li could somehow mimic a mussel's glue in the lab, eliminate the unwanted solidification, and scale it to amounts that are useful for industrial applications... well, then he would have something of value. But to do so, he needed to add one missing ingredient.

Soy What?

TO MOST OF us, soy is known as an ingredient to add to stir-fries, or the source of a milk that makes our lattes creamy, or maybe the base of an eco-friendly candle. However, soy also has a history as a glue. Soy glue was common between the 1930s and the 1960s, before stronger, more water-resistant adhesives superseded its reign. Unlike formaldehyde, which is naturally limited and made from petroleum, soy protein is plant based, annually renewable, and nontoxic. When it comes to building a better glue, then, soy is pretty appealing.

Li certainly thought so. But in order to compete with other glues on the market, he needed to make a soy glue that was stronger than the usual variety—much stronger. And so, he took the bonding might of mussels, with their L-DOPA superpower, coaxed it together with soy, and made a glue with the strength, water resistance, and cost-effectiveness needed to compete with other resins on the market. The soy-mussel super-duo gave him the best of both worlds. Instead of creating mussel proteins from scratch, Li's team modified the soy, blocking proteins the mussels lack while transferring some of the mussel's amino acids into the soy complex. The result is a soy protein imbued with a Frankenstein mix of mussel strength. Even after hours of boiling, the glue doesn't degrade.

Li proved his idea could stick in the lab. Next, he needed to conquer the market. This is the point at which science often fails to make it out of the lab's pipettes and petri dishes. What gets in the way is a double-barreled challenge: society's general resistance to change, and industry's unwillingness to spend money on innovations that might not pan out. Li needed backing from a company interested enough in his idea to upend tradition and make a dent in the adhesive industry. In 2003, that opportunity came when he met Steve Pung, vice president of technology and innovation at Columbia Forest Products. Established in 1957, Columbia Forest Products is the largest manufacturer of hardwood plywood and veneer products in North America. After the World Health Organization formally recognized formaldehyde as a potential human carcinogen, the company was on the hunt for a new, formaldehyde-free resin.

They snagged Li, and a resin supplier for the paper industry soon joined the team. Columbia Forest Products revised its manufacturing process for hardwood plywood panels to incorporate its new technology based on Li's innovation, called PureBond, replacing an estimated 47 million pounds of conventional formaldehyde resins and reducing hazardous air pollutants by 50 to 90 percent from the company's products. This made Columbia Forest Products the first company in the wood production industry to manufacture a cost-competitive formaldehyde-free, soy-based resin. Of course, this doesn't make the production of the glue net-zero in carbon emissions; the glue is simply a step toward a less toxic, renewable option. And while it's always possible that a newcomer on the market may be worse than its predecessor, soy adhesives don't present a high risk for health concerns. Although soy is a potential allergen for some (mostly infants), causing digestive, respiratory, and skin reactions, the extensive processing during the manufacturing of the glue likely removes these compounds. Some in the field are calling Li's adhesive an "ingenious chemical construct, something of a Holy Grail."[60]

Less than three years after Oregon State University patented the adhesive formulas, Columbia converted its seven plants to the nontoxic soy-based glues. In July 2019, Columbia announced it had manufactured its 100 millionth PureBond plywood panel using adhesives from American-grown soybeans.[61] Li's glue has now been adopted in some 60 percent of American plywood products, according to Oregon State University.[62] For his work, Li was awarded the 2007 Presidential Green Chemistry Challenge Award by the Environmental Protection Agency.

Of course, Li and Waite are not the only scientists curious about these sticky bivalves...

Sustainable Engineers

MARINE BIOLOGIST EMILY CARRINGTON stands in front of a class of students and holds out a murky jar filled with so much phytoplankton you can't see through it. She plops a mussel inside, puts the jar aside, "and in less than an hour you've got crystal-clear water," she says.

Carrington is a professor at the University of Washington who has studied mussels for decades, first on the East Coast and then the West Coast just south of Vancouver on San Juan Island. "We've got a lot of agricultural runoff in our estuaries," she says, "and all that nitrogen we're pumping out is driving a lot of the productivity in the estuaries, which can be good and it can be a problem when there's too much. My fantasy is that by putting a curtain of mussels in front of that outflow, the mussels will clean it."[63] That "would be so cool," adds Carrington. You remove the mussels after they've been harvested and use them back on the land as fertilizer. "We're helping to close the loop. Right now, all the nitrogen is a one-way ticket to the ocean."

Mussels are, as *Salish Magazine* calls them, marvelous "janitors of the ocean."[64] An individual mussel filters a bathtub's worth of seawater every day, drawing it through a siphon on one end and expelling it out the other. They filter-feed on microscopic sea creatures and, in the process, remove biological and synthetic toxins from the water. These toxins then either are broken down or accumulate in their soft tissue. With this in mind, the Washington State Departments of Ecology and Fish & Wildlife are now using mussels as measuring sticks for contamination in the nearshore habitats of Puget Sound. The most common contaminants found in the mussels' guts are pollutants from crude oil, gasoline, burned garbage, flame retardants, electrical cables, pesticide extenders, and insecticides. The levels of these pollutants rise near the sound's most urbanized areas, including marinas and ferry terminals. Given that mussels store much of what they filter, it's no surprise that there are strict regulations on what can and cannot be eaten.

A Stanford University study showed that mussels even remove and inactivate *E. coli*—an indicator of fecal contamination—from the water. Another concern is the runoff of fertilizers and sewage into our waters, both of which contain phosphates and nitrates that cause algae to proliferate, starving the water of oxygen and creating "dead zones." Mussels can recycle these nutrients when they feed on excess algae. Time is, however, of the essence.

Carrington's team combined laboratory experiments with a mathematical model to show that high carbon dioxide concentrations can weaken the mussel's super threads, dislodging them at current wind and wave forces. Carrington says these threads are like our hair. When nutritionally stressed, our hair falls out; the same happens with mussel threads when the water slips closer to acidity on the pH scale. This future is nearer in Puget Sound than in much of the rest of the country due to upwelling, where cold, nutrient-rich waters rise to the surface.

"That deep water is kind of skanky; it has a lot of CO_2," says Carrington. This makes Puget Sound's waters a hundred years ahead of the curve when it comes to climate change–driven pH levels. For mussel farmers at Penn Cove, two hours north of Seattle, Washington, this is also a problem. Penn Cove has some of the oldest commercial mussel farms in North America. Harvesters splash thick ropes into the water that are attached to massive wooden rafts, and they haul the ropes back up after about fourteen months. The harvesters never need to feed or fertilize the mussels; they only give them a good place to thrive. Each of their rafts supports between 900 and 2,500 mussel lines, yielding up to fifty pounds of mussels per line. The company harvests over 2 million pounds of mussels per year from their two farms in Washington State. But if the threads start to weaken, the mussels are more likely to fall off when the rope is hauled up. "It's kind of like a fruit falling off early from a tree before you've had time to collect it," says Carrington.

This is an unfortunate turn of events. We have yet to fully realize the mussel's potential, especially when it comes to biomedical uses. Scientists are looking into the hidden complexities of mussel glue to inspire the next generation of self-healing hydrogels and surgical wound closures. Phillip Messersmith, a professor at UC Berkeley, took cues from mussels to make a surgical glue that can seal the delicate sack a baby grows in in the womb. He has partnered with Michael Harrison, the "father of fetal surgery," who performed the first open fetal surgery three decades ago at the University of California, San Francisco. Stitches are a decent means of sealing up simple wounds, but when suturing tissues inside the body, it is sometimes better to glue than to sew. Sutures and staples have a higher risk of patient discomfort and infection than sealants because they create small damage to the surrounding tissue. However, most adhesives don't work well in a moist environment. Mussels, on the other hand, have evolved over millions of years a glue designed for a wet, salty habitat; fetuses live for nine months in a wet, salty environment. Of course, there are differences too, which is why research is underway to see if there is enough oomph behind the idea.

Other mussel species have invented different glues using chemical artistry fit to stir the heart of a scientist. Some species glue themselves tight to seagrass, others to rocks near deep-sea hydrothermal vents. Emerald-green mussels in Singapore use different techniques from California mussels because their waters are hot and oxygen rich. These green mussels have a greater challenge to overcome than the California mussels in preventing oxidation. "That's a wonderful model system attribute to find because oxidation is a problem with all systems," adds Waite, "and to find adaptations that are tuned to the degree of oxidation possible allows you a much more comparative picture when you're developing a product or even a model for how the mussels do it."

Across brown mussels, green mussels, and Siberian mussels, the triggers and curing mechanisms for these bivalves are slightly different. "The rub is that [these are some] of the most wonderful modeled systems for studying different processes and different approaches, yet they're at risk of extinction," says Waite. "They won't be around to study."

Many of the more than three hundred mussel species in the United States are endangered, with more waiting to be added to the list, including the wonderfully named orangefoot pimpleback, fat pocketbook, purple cat's paw pearlymussel, rough pigtoe, and sheepnose.

Making It Stick

WHEN IT COMES to sticky stuff, mussels don't have a corner on the market; nature is a workshop of gluey substances. The carnivorous cape sundew plant lures insects to its red tentacles and ensnares them in sticky pearls of temptation. The sea cucumber shoots out a white spaghetti-like substance from its back end to tangle an adversary. Snails use a slimy foot to slide up walls. Spiders spin sticky webs to catch prey while their own spindly legs remain immune as they tightrope-walk across the lines. Cape rain frogs in South Africa have taken the adage "sticking together" to the extreme. Their bodies are so round and their legs so small that they couldn't possibly hold on to each other when they mate. Instead, they ooze gluey stuff from their skin to help them stay close.

Caddisfly larvae are perhaps the most peculiar, gluing pebbles together in rivers, marshes, and tumultuous streams using their sticky spit. Have you ever tried to stick a bandage on in the bathtub? If so, you know it's not easy to glue things underwater, and yet, in a river, the slobber of caddisflies sticks better than tape. As they grow, these natural-born builders use their black twiggy legs to search for just the right pieces for their homemade homes. The larvae bind together not just pebbles but sand grains, twigs, and leaves. The elastic properties of their glue allow it to stretch and absorb physical force while also not being so bendy that it snaps back like a rubber band. Researchers are analyzing its chemistry in the hopes of making a synthetic version that can be used as a surgical adhesive.

The stickiest organism of all, however, may just be a bacterium (*Caulobacter crescentus*). The edamame-shaped bacterium, found in both freshwater and seawater habitats, is so sticky that just one square centimeter of its natural glue can support 1,500 pounds (or the weight of a bison) to a wet surface. The bacteria swiftly colonize boat hulls, water pipes, and medical catheters, and are notoriously hard to remove. In 2006, a team of scientists tested its strength and found that it requires a force of 70 newtons per square millimeter to remove the bacterium from a surface. That makes it almost three times as strong as commercial superglue.

Sticky stuff, however, is not nature's only option when it comes to keeping things together. The burr is a famous example; in fact, it inspired the invention of the Velcro hook-and-loop fastener. George de Mestral, a Swiss electrical engineer, devised the fastener after a hunting trip in 1941, when he noticed burrs stuck to both his trousers and his Irish pointer's fur. The two of them were covered in the things. They were a nuisance, an annoyance, buggers to constantly pluck off his pants. When he returned home, he put a burr under a microscope and saw thousands of tiny hooks. Perhaps, he wondered, we could use the stuff to fasten clothing rather than using buttons or zippers. Inspired, he worked by hand to create one piece of cloth that had tiny hooks similar to those he saw on the burr, and another piece of cloth with loops. He patented his invention and named the product Velcro, from the French words for velvet (*velour*) and hook (*crochet*).

In biographies of de Mestral, his invention is most often attributed to serendipity and luck. One book even calls him "Lucky George." But there's so much more to his story than that. At some point in our lives, we've all been plagued by something or another. De Mestral was irritated by burrs, and he decided to take a closer look. It wasn't luck. It was curiosity and grit. He proved he could create a burrlike fastener, but mass manufacturing was another conundrum. He took his idea to six fabric companies and they all turned him down. With his money running out, he retreated to a tiny cabin in the village of Commugny in the Swiss Alps and eventually came up with a device to make the loops and fasteners more quickly. Nearly twenty years after having his initial idea, he was finally able to mass-produce the product.

Velcro's hook-and-loop fastener is now used all over the world. In the 1960s, Apollo astronauts used Velcro to secure objects to the walls in order to prevent them from floating away. Today, Velcro is used in everything from hospital blood pressure cuffs to car floor mats to shoes.

A Battery-Operated Adhesive

NO DOUBT ABOUT it, glue has worked wonders for society. In the age of synthetic chemistry, though, the question on the table is "Can we do better?" It's not an easy task. We often require our adhesives to take on properties and encounter scenarios not typically seen in nature. Aircrafts use adhesives, but there are no massive metal birds in the sky stuck together with glue from which we can take direct inspiration. Instead, we consider chemistry and contemplate whether there are useful ideas we can incorporate into our daily needs.

Waite's new interest is the chemistry of a "living" adhesive. When the mussel's foot lifts off the surface of a rock, L-DOPA, which helps the mussel to stick to rock, is in great danger of oxidation. But—and this is what excites Waite—the L-DOPA that resides in droplets during its cellular storage is shielded from oxidation. "So what?" you ask. So, when L-DOPA gets oxidized, the shielded L-DOPA can, by a battery-like association with the surface film, provide the oxidized L-DOPA with electrons and protons to return it to its pristine state. "This is a dynamic association that takes place between the chemistry at the interface and the chemistry in the architecture above the interface. So it's really a living adhesive," says Waite. "As it grows old, the L-DOPA that's invested in the droplets where it's protected from oxidation restores the youthfulness of the adhesive layer until the mussel plaque runs out of that supply. So that is like a battery running down."

The no-longer-useful components drop from the mussel and continue to feed the lowest life forms in intertidal zones, just like the leaves of a tree eventually die and fall to the ground to become important nutrients to feed another cycle. "This is beautifully tied in with how nature engineers things not to be immortal," says Waite.

"So, yeah, a battery-operated adhesive," he says with an excited grin, his enthusiasm mirroring that of many scientists reveling in the unknown of what other wondrous delights are crawling, growing, or gluing out there in the intertidal zone.

7

CONCRETE EVIDENCE

Coral-Inspired Cement Drastically Reduces Construction Industry Emissions

"It's our power to invent that makes me hopeful."

BILL GATES

THE BIRTH OF a new building is like the delivery day of a baby—at least in some regards. When the new entity is welcomed into the world, oversized scissors cut a symbolic umbilical cord and cheers and smiles erupt all around. Cameras click to capture the moment, and it's not out of the ordinary for the ritual pop of a champagne bottle to be heard. It's a momentous event, months in the making; it represents a new phase in someone's dream. And yet, unlike with the birth of a child, this is where the excitement usually ends.

Soon enough, the cut ribbon is torn down and thrown in the trash, the massive scissors are shoved in a drawer, never to be used again, and the empty champagne bottle is tossed in the recycling bin. There are no plans for the future of this "concrete child," and no thought at all as to the fate of its materials in the decades to come. For the longest time, we ignored the construction industry in discussions about climate change, focusing instead on electric cars and unplugging our appliances when they are not in use. Now, the construction sector has been dubbed the "last mile" of decarbonization. Buildings are responsible for some 40 percent of the world's emissions, according to the International Energy Agency, and yet we've slammed into a brick wall when it comes to designing low-carbon and net-zero buildings. To reach those goals, everything involved in the construction (birth) and maintenance (life) of a building must be taken into account and decarbonized. It's a formidable challenge, especially considering the only thing humanity consumes more of by volume than water is concrete.

Gray, lifeless, and boring, concrete is the world's most widely used human-made material. Look outside your window. Can you see a street curb? It's likely concrete. The tallest skyscraper in the world? Again, concrete. The Pantheon in Rome is the largest unreinforced concrete dome in the world, and it's been standing for nearly 1,900 years. Concrete is so ubiquitous that it's even entered our lexicon to mean something that is certain or existing in a form that can be seen or felt. There is "concrete evidence," and someone has "concrete ideas." Concrete is under our wheels on highways as we drive bumper to bumper to our jobs, homes, and dreams. It is the bones of our cities, used in everything from buildings and bridges to pipelines, drains, and dams. It is both a lifeless building block and a lifeline of humanity, the foundation of globalization and a contribution to our possible downfall. As a material, it is resistant to rust, nonflammable, durable, and inexpensive, and it doesn't rot.

And yet, concrete is also one of the world's leading producers of greenhouse gas, releasing up to 8 percent of the world's carbon dioxide emissions; the planet's concrete emissions are greater than any single country's alone (save China or the U.S.). Global concrete production in 2020 generated a $617 billion industry, according to Allied Market Research, and still, in many regions it remains one of the least regulated activities. "Sand mafias" try to cash in on the global enterprise, illegally dredging up and bagging sand, a key ingredient of concrete.

By the time today's babies reach middle age, the number of buildings on Earth will have doubled, according to projections in a UN environment report. That's the equivalent of one New York City every month. And like so many of our existing buildings, many of these new ones will be made of concrete. Concrete can hold thousands of tons, doesn't get infested with termites, and is easy to use. Just mix a batch and pour your foundation; the job can be done in a day. It took about eighteen hours to pour the base of the tallest building in San Francisco, a concrete slab 14 feet thick, nearly an acre wide, and containing 49 million pounds of concrete. At the time, it was one of the longest concrete pours in history, only to be surpassed a couple of years later in Dubai with a residential commercial building.

Concrete's explosion in modern industry has reached a colossal, unprecedented scale. In many ways, it is still largely the same industry it was at its humble beginnings. But a radical new dawn is upon us.

It's Alive!

I'M CRAMMED INSIDE a pitch-black room with enough space to fit myself and another person. That's it. There is no light. It is can't-see dark. Except. Click. A red flashlight flickers on.

That's when we see an aquarium of coral. It all seems a bit over the top for this undercover containment facility, but the scientists here are trying to reproduce coral, a notoriously tricky task. The coral are hermaphrodites, meaning they possess both sperm and eggs. One would think this would make it easy to get them to procreate, but it turns out coral are also underwater romantics. They require just the right conditions to "get them going": they favor a perfect water temperature and will spawn only after the flush of light of a full or new moon. If the ambience is seductive enough, the coral synchronize their spawning on the same matchmaking night and release their bundle of eggs and sperm (which resemble small seeds), and the buoyant bounty floats to the surface.

When I visited San Francisco's Steinhart Aquarium a few years ago, few institutions had managed to make coral release a cloud of sperm in the lab. "It's like snowfall in reverse," said Dr. Bart Shepherd, director of the aquarium, as we stood there awkwardly looking at the dual-sex coral doing nothing remotely sexy.[65] When I looked online to see what he was talking about, it became clear that his description was spot on. The orgiastic blizzard rises to the water's surface like trillions of little pastel-pink balloons. Finally, in April 2020, the Steinhart Aquarium made it happen. If the "night of action" had occurred in the ocean, the free-swimming larvae would have gone for a ride, floating for days or even weeks before they found their forever homes. The larvae use chemical cues to guide themselves toward a reef and then drop to the ocean floor. The equivalent for us would be if someone else's child randomly descended from above and started growing out of our side, except on a reef everyone is connected and sharing nutrients. Because of this connection, Earth's largest living structure is the Great Barrier Reef, reaching 1,400 miles (2,300 kilometers) in length, or almost the vertical length of the United States. In fact, the maze of coral is so grand, astronauts can see it from space.

It's these kinds of coral that Ginger Krieg Dosier cares about. In photographs, Dosier typically wears black and is often seen holding a concrete brick. Not surprising, given her job. Dosier is the president and chief executive officer of Biomason, a biotechnology start-up in Raleigh, North Carolina, that grows concrete bricks like a gardener grows carrots or cucumbers. Well, not quite. A better analogy is that Biomason's bricks grow like hard coral reefs in the ocean.

Dosier stumbled upon her inspiration for homegrown bricks on the sandy beaches she played on as a child, picking up seashells and wondering how they grew. At the time it was nothing more than childhood curiosity and games. It wasn't until she was working toward a master's degree in architecture that her interests in design and the ocean merged into an attempt to upend a centuries-old industry. And it all started with a small lab she fashioned in her kitchen, which soon spread to the second bedroom of the apartment she shared with her partner, Michael Dosier. In their version of romance, they'd often return early from their dates to "feed" the cement with different strains of bacteria, testing out dozens to unearth which binds best to the sand grains used in cement. The Dosiers were trying to replicate how coral grow in the sea in their homemade laboratory.

Coral may seem like lifeless towers at the bottom of an otherworldly domain, but they are as much an animal as a tiger or an elephant. Scale and complexity are the wondrous might of the animal kingdom and, until recently, coral have been amorous superstars. There are more than 3,600 species of hard coral, their intricate architecture ranging from fingerlike protrusions to flat disks to squiggly bulges that look as if thousand-pound brains have ruptured from the seafloor. Just like humans, coral exhale carbon dioxide; unlike us, the coral's CO_2 combines with calcium in the ocean to create calcium carbonate, the chalky compound out of which their skeletons are made. Coral reefs are made of hundreds of thousands of tiny builders called polyps, each the size of a pinhead. These flower-shaped polyps convert sunlight into energy to form a coral colony. To grow, coral polyps push ever so slightly away from a surface (on rocks or on top of each other) and create a gap; they then fill that gap with calcium carbonate. With no central brain, a coral colony acts as a single organism, its stinging cells and tentacles catching prey that drifts by. Each polyp in the colony grows at a rate of between half an inch and seven inches each year, depending on the species. Over time, these colonies merge into a reef and—again like us—their skeletons can persist after death for thousands of years, their dead remains scrolls to past oceanic conditions.

Coral skeletons add annual layers like trees add rings, each layer a time capsule of the water's temperature, pollution, and geology. Coral skeletons even contain ocean records of nuclear weapons tests during the 1950s and 1960s. To extract a coral core, a diver drills a long hole with a circular blade, the dust from the coral's skeleton clouding the waters, and then seals the hole with—you guessed it—concrete. The coral will then slowly grow over the dead gray stuff; that is, if global warming doesn't drain them of color first.

A reef's polychromatic splendor is created by millions of algae living inside the coral's clear tissues. More than 1,600 species of "living pigment" paint coral reefs flashy pink, tangerine orange, vivid violet, lemony yellow, or pickle green. Some even glow. Think of the coral as the canvas, the algae as the pigment, and the water as the painter. This symbiosis is coral's greatest strength and also its Achilles' heel, making it sensitive to a warming climate. During heat waves, overheated polyps perceive the algae as an irritant and spew them out, revealing their stark white skeletons. The coral are then left bleached and starving.

The symbiotic relationship between algae and coral began some 210 million years ago, during a time when coral quickly spread across the globe. The blueprint for Biomason's design has been underwater all this time. The bricks are grown by speeding up what happens in nature over thousands of years into a few days in the lab. Sand packed into rectangular molds is injected with bacteria (*Sporosarcina pasteurii*), while a sprinkler provides the nutrient-rich water the bricks-in-waiting need to grow and harden. The sand acts as nuclei for the bacteria, which wrap themselves around the grains and form calcium carbonate crystals, creating a bio-cement between the sand grains and strengthening the brick from within. After three to five days, the bricks are strong enough to be used in buildings. To make sure the entire process is as green as possible, the water is recycled for the next batch. The bricks are made of approximately 85 percent granite from recycled sources and 15 percent biologically grown limestone.

The study of how to make these bio-bricks is part of a growing field called engineered living materials, where live organisms are added to something not alive to create a new material. It sounds absurd at first, like something from Mary Shelley's *Frankenstein*, but in fact we already make good use of microbial cell factories in our day-to-day lives. The bacterium *Propionibacterium*, for example, is responsible for the holes in Swiss cheese. When the bacterium gobbles up the lactic acid in the dairy product, it releases carbon dioxide as a by-product and forms tiny bubbles in the wedge. In the pharmaceutical industry, bacteria are used to produce antibiotics and vaccines. Many forms of bacteria are our allies, not our adversaries.

But traditional concrete is nothing like these examples. It is a mixture of water, aggregate (rock, sand, or gravel), and cement, the last of which is a binding agent that holds it all together. Those in the industry like to say that concrete is the cake, or final product, and cement is the flour. To make Portland cement—the industry standard today—the first step is the mining of the raw materials, primarily clay minerals and limestone. This is dumped into an impact crusher machine and pulverized, and then transported on conveyor belts to a factory where it is roasted in a huge kiln at temperatures of up to 2,700°F (1,500°C). Crush in some gypsum (a mineral with a white luster found in layers of sedimentary rock) and, voilà, you have Portland cement.

So where do all of those carbon dioxide emissions come from? Cement is typically made with limestone because of its high levels of calcium (plus carbon and oxygen). Limestone is what remains of countless generations of shelled marine creatures such as coral and shellfish. When these creatures die, their shells and skeletons accumulate as a sediment. Over millions of years, powerful geologic forces compact this sediment, leaving what we see when we look at limestone today: a chalky white rock. Pick up a seashell and you are holding calcium carbonate, the raw material of limestone. Calcium is a vital ingredient in cement, but the limestone's carbon and oxygen isn't needed, so we burn the rock to get what we want: all the calcium. The chemical reaction that rids limestone of its carbon and oxygen creates the carbon dioxide that enters our atmosphere. It's a simple but powerful one-to-one ratio: the creation of a ton of cement releases around a ton of carbon dioxide.

According to the United Nations Environment Programme, enough concrete was created in 2012 alone to construct a wall 89 feet high by 89 feet wide (27 by 27 meters) around the equator. That's a lot of concrete—and a lot of CO_2. So what can we do about all the carbon floating in our atmosphere?

Capturing a Ghost

HOW DO YOU capture a ghost? A colorless, odorless gas? The idea sounds almost whimsical, especially when the ghost in question is composed of carbon dioxide, an eerie threat looking us right in the face. The carbon portion of carbon dioxide is not inherently bad. Without carbon, Earth's ocean would be frozen solid, an icy wasteland. In fact, all known life on Earth is based on carbon. Yet, carbon is also a promiscuous element; it likes to bind to other elements to form new compounds. When carbon finds two oxygen atoms for a three-way love affair, the result is carbon dioxide—and that's when things get tricky. Too many of these three-way liaisons intermingling and banging into each other make Earth too hot to handle. The heat they trap from the sun warms the planet. One solution is to capture the offender, to grab it out of the air. But to snatch something with no fixed shape or volume—in essence, a gas—we need to get creative.

Carbon capture technology has been around for decades. Perhaps the most well-known example can be found on the International Space Station. When astronauts exhale on the space station, the carbon dioxide has nowhere to go; it is bound by the station's aluminum alloy walls. Without fans to circulate the air, pockets of carbon dioxide can build up in the cabin and form a cloud around the astronauts' heads. Carbon dioxide concentrations above 10 percent can cause humans to convulse, slip into a coma, and die. Astronauts on the space station cannot simply open a window to get more fresh air. This is why they always sleep next to ventilator fans; without them, they might suffocate as they sleep. The Ames Research Center has noted that metabolic wastes (expired carbon dioxide, evaporated sweat, urine, utility water, and feces) add up to ten to fourteen pounds per person per day to the station. For a three-year mission with ten astronauts, that's more than 100,000 pounds—without considering the containers needed to store it all. So how do they do it?

Engineers have devised a removal system that scrubs the air of carbon dioxide, then dumps the CO_2 overboard into the vacuum of space. The system houses synthetic rock beads with tiny pores called zeolite—think of something like a sponge that's been turned to stone. Carbon dioxide sticks to the zeolite, while other molecules zoom right through. Space shuttles also use filtration canisters to absorb carbon dioxide using a chemical reaction of lithium hydroxide, but after about a month, the canisters become less cost-effective than the zeolite system. Such methods are viable for six or so people on a closed space station, but Earth is much larger—and has much more carbon dioxide to capture. The gas is also more dilute in our atmosphere, meaning filtration canisters would need to be produced at a scale that isn't yet feasible and the CO_2 would have to be buried underground. Scientists need to devise new ways to take control of this toxic specter in our midst.

If we were to stare at Earth from the window of an aircraft today, we wouldn't see carbon dioxide at all. But we could infer from the sprawling cities below us that a creature with a mind-boggling talent for construction lives down there. The view stands in stark contrast to what we would have seen from the same vantage point thousands of years ago, when there was no concrete. There weren't even moving trucks or pack animals for ancient foragers. They carried everything they owned on their backs. Until relatively recently, most people lived frugal, modest lives. With the exception of the rich minority of nobles and royalty, people were sparing with their things, sewing up torn sheets and shying away from buying two of the same item when one would clearly do. These days, though, just one middle-class person likely owns thousands of things, from homes and cars to curtains, pens, and blenders. For a collective of billions, this adds up to lots of stuff, whether that be the number of shampoo bottles sold per year (548 million in the U.S. alone) or the 1 billion toothbrushes in the U.S. thrown in the trash each year. Then there are humanity's collective pleasures, like soccer stadiums (nearly 5,000 worldwide) and Catholic churches (more than 220,000). Even our stuff takes a lot of stuff to make. Churning out plastic toothbrush after plastic toothbrush requires industrial-scale machinery. And to make that machinery? You guessed it: more stuff.

We've perpetuated a culture where stuff begets more stuff and that stuff can start to take a physical and mental toll. Is it the love of stuff that's the problem, or is it not loving our things enough, perpetuating a culture of disposable goods? To swell with stuff is arguably just as unsustainable as to grow meager with too little. Who isn't tempted by a potted plant that livens a room or a comfy bed to rest on at night? The question then returns to: What if we consider the life of the things we value? That to give it a space in society is enough motivation to also provide it with an end-of-life plan?

Unprecedented consumption has ticked our emissions higher and higher. In the 1980s, human consumption exceeded our planet's regenerative capacity for the first time. One year before the dawn of a new millennium, the carrying capacity of humanity swelled to 120 percent, meaning 1.2 Earths were needed to regenerate what we used up. Concrete is a huge player in this statistic. To construct a net-zero building with no carbon footprint, we need to fundamentally rethink its design, from its creation to how it's used. "Embodied carbon" is a term used to encapsulate all the emissions generated from the extraction of materials, manufacturing, transportation, and assembly of a building, which comprises about half of the building's carbon footprint. "Operational carbon," or the emissions when the building is in use, makes up the other half. Together, embodied and operational carbon account for 10 percent of the planet's annual emissions.

Here's where it gets tricky. There isn't one kind of concrete that's used for everything. There is high-density concrete (atomic power plants), prestressed concrete embedded with reinforcement bars (bridge construction), rapid-hardening concrete (underwater projects), asphalt concrete (highways), pumped concrete (high-rises), concrete with accelerators or retarders to speed or delay hardening, and many more. Not only that, but there are key differences between the way nature operates and the way humanity does: for one, nature's work is fleeting, a perishable art where the chemistry of one creature that expires mixes to become part of a tree or the energy of another critter. Flip back to us, and most of our stuff piles up in heaps that we ship off to bigger heaps. Dumped concrete takes centuries, or even millennia, to crumble back into sand. Humans are not yet so sophisticated to accomplish nature's circular feat, though we try with primitive versions like recycling. When we construct buildings, we design them to last. Of course, this isn't entirely the case, as Alan Weisman's book *The World Without Us* reveals. If humans were wiped from the face of the Earth, plants would creep over our cement and take over. It's our constant "pruning" of concrete creations that keeps them standing in gray glory.

The twenty-first century will likely see a paradigm shift in how buildings are made. "The conditions have never been more clear for backing energy breakthroughs," says Microsoft founder Bill Gates, who has turned his attention, and wealth, to helping solve the climate crisis.[66]

Growing the Future

GLOW-IN-THE-DARK CONCRETE. RED brick-like concrete. Concrete with a glossy finish. Concrete with hidden drawings that become visible in the rain (hello, Seattle!). These are all well within Biomason's toolbox. Where traditional cement frees the carbon contained in mined limestone using extreme heating, thus shedding carbon dioxide as a by-product, Biomason's bio-cement uses no heat at all. In fact, the process is rather simple. Sand and bacteria are placed into a mold, fed water with calcium ions, and squished down with a pressurized plate. Similar to the way coral grows, the bacteria create bridges of calcium carbonate between the grains of sand. Biomason is essentially letting the bacteria do the burdensome job of making calcium carbonate for them. This means their bioLITH tile releases 99.4 percent less carbon dioxide than a regular concrete block and is three times stronger.

Dropbox used the bio-bricks for an exterior courtyard at its headquarters. Martin Marietta, a supplier of aggregates and heavy building materials, has also installed the tiles outside its headquarters. In 2021, Ginger Krieg Dosier's company signed a deal with a top clothing retailer to create low-carbon flooring. With support from the United States Department of Defense, the company is developing a bio-cement tile that is agile and can be deployed on non-engineered surfaces. We'll soon see if the company can succeed on an industrial scale: time, speed, and money are the currency of industry, although it's perhaps worth considering that we may need to slow down, to recognize that speed at the cost of atmospheric destruction is not the price we want to pay.

In the last five years, fledgling cement companies have sprung forth to compete in an intellectual arena of the fittest. They all have their eyes on a central theme: capture or emit less CO_2. CarbonCure sources carbon dioxide for cement from industrial emitters, which collect, purify, and distribute the gas in pressurized tanks to concrete plants. The company's technology then injects the captured CO_2 into the wet concrete mix. The carbon dioxide reacts with the calcium ions in the cement to form calcium carbonate, making the concrete stronger, just like when coral release carbon dioxide and their "breath" mixes with the calcium in the ocean to construct their skeletons.

CarbonCure's technology is a two-part device that is installed at an existing concrete plant. The first piece of equipment is a computer that sits in a control room. When the concrete producer says, "Okay, I need another truckload of concrete, and it has to be this mix design," CarbonCure's computer communicates with the pressurized tank of carbon dioxide, where the second part of the equipment sits. This piece is a valve box that injects a precise dosage of carbon dioxide at the right time into the wet concrete. CarbonCure is currently used in nearly three hundred ready-mix plants around the world and was recently backed by Bill Gates. LinkedIn also used this technology in the building of its 245,000-square-foot headquarters in Mountain View, California. Theoretically, CarbonCure could capture CO_2 from the cement industry itself and reinject it back into the cement at a later stage in its processing, but for now that's a hope for the future.

Other players trying to decarbonize the cement industry include Blue Planet, Solidia, and Carbon Engineering, to name just a few. Blue Planet, whose founding entrepreneur was also inspired by coral reefs, uses carbon dioxide collected from the exhaust stacks of power plants as raw material. Calcium sourced from waste like demolished concrete, kiln dust, or fly ash is then combined to manufacture synthetic limestone. There's no purification step, which they say is an energy- and capital-intensive process. Blue Planet's aggregate was used at San Francisco International Airport in 2016.

Solidia, for its part, produces its cement in traditional kilns by curing concrete with carbon dioxide, except the company uses more silica and less calcium carbonate, so the kiln can operate at temperatures 480°F (250°C) lower than the norm. This cuts emissions by at least 30 percent, Solidia says, and up to 70 percent in certain cases. Solidia technology is used at more than fifty concrete manufacturing facilities in ten countries.

According to Bill Gates, these companies are taking steps in the right direction, but more needs to be done. The fate of concrete is inextricably woven with ours—a simple substance with a complicated future in the face of a changing climate. "Setting a goal to only reduce our emissions—but not eliminate them—won't do it," he writes. "The only sensible goal is zero."[67]

8

DRIVING ON A SEED

Pomegranates and Abalone Inspire the Next Generation of Batteries

"A scientist is inevitably addicted to the unknown."
GREGORY MOISEYEVICH LEVIN, Soviet botanist

A JUICY-RED TESLA swerves in and out of Silicon Valley traffic. Yi Cui, a materials scientist at Stanford University, is at the helm, driving to work at his company, Amprius. *Science* magazine is along for the ride to chronicle his investigations into a new world of lithium-ion batteries. Cui's is not the first electric vehicle on the road in 2016—roughly 2 million join him around the world. Electric cars are widely heralded as an improvement over gas-powered vehicles, cutting down on the use of fossil fuels, but they're not enough to save us from ourselves on the climate front. Cui knows this, and is looking to do better. Where computer chips have surged in performance over the years, the storage capacity of batteries has, by comparison, trudged along at an unimpressive rate. A single charge on a Tesla takes the car about 260 miles (400 kilometers). Consumers are looking for better, and manufacturers want to deliver. But to make it happen, we need better batteries. After years of trying various iterations, Cui's team found the juice they needed in the most unlikely of places: pomegranates.

In ancient Greek mythology, pomegranates were known as the "fruit of the dead," sprung from the blood of Adonis. Crack open a pomegranate and it looks like a broken heart held in soiled hands. The fruit's mythology is rife with other metaphors, too, often around fertility and rebirth, but also of possibilities and futures. The ruby globes have been cultivated since ancient times and are featured in Egyptian mythology, the Old Testament of the Bible, and the Babylonian Talmud. Each pomegranate is a symbolic ovary, the essence of fecundity, its leathery orb protecting hundreds of garnet seeds. Some even believe Eve plucked a forbidden pomegranate from the Garden of Eden.

The pomegranate may seem an unorthodox source for new battery ideas, but it's not the first time batteries have benefited from an unlikely inspiration. The birth of batteries can be traced back to the 1780s, when Luigi Galvani suspended skinned frog legs to a brass hook. Galvani was an Italian physician, physicist, biologist, and philosopher at a time when a person could choose to pursue many professions at once. He's frequently pictured wearing a paper-white wig with tight barrel curls, as was typical of the style in the eighteenth century. (The fashion died out because the wigs were made from goat and horse hair, smelled quite terrible, and attracted lice.) In an experiment that shocked the world and changed it forever, Galvani touched the blade of a scalpel to a dead frog's leg and the leg unexpectedly twitched. Yep, that's all it took to change the world forever. For years, scholars debated whether electricity in the tissue was responsible for the twitch (as Galvani believed) or whether the dead amphibian was simply transmitting a current that flowed between the two metals (brass hook and steel scalpel). Scientist Alessandro Volta eventually proved it was the metals, not the frog's tissues, that produced the current. In 1800, he made the revolutionary "voltaic pile," otherwise known as the world's first battery. He stacked alternating disks of copper and zinc, which were separated by a cloth soaked in salty water. Wires on both ends produced a stable current.

At the time, batteries seemed like alchemy, and undertones to that effect are present in the early years of exploring their potential. Energy is often a hard-to-see phenomenon, and even now many of us have only the vaguest notion of how batteries work. Perhaps more evocative is the fact that few of us *expect* to understand, so let's quickly review the basics. Electrons are tiny particles with a negative charge. Electricity is the stream of electrons. Volta's battery works because the copper creates a positive pole (known as a cathode) and the zinc forms a negative pole (anode). The pile doesn't have a charge until something conducts electricity, like the metal wires connected between the two poles. When this happens, electrons flow from the negative pole to the positive to create a battery, albeit a primitive version.

Nowadays, batteries are a superpower, giving us the ability to harness energy and store it for future use. If you're camping in the woods at night, you can insert a AAA battery into your flashlight and shine a light by converting chemical energy into electrical energy. Rechargeable batteries like those in cell phones and laptops are even better because we can plug them into an electrical wall outlet when they lose their charge. By applying a reverse current, we make the electrons flow back to their starting position in the anode.

But making better batteries isn't just about cramming as many loose electrons into as small a space as possible. An electron needs to be freed from an atom, which comes with baggage, specifically protons and neutrons, both of which weigh 1,800 times more than an electron. One difference between a lead-acid battery and a lithium battery is the amount of baggage they carry. Electrons in a lead-acid battery come with 82 protons and 125 neutrons. Lithium comes with only three protons and four neutrons, making it substantially lighter. In fact, at the start of the automobile age, electric cars were almost our future, not gasoline. To explore why, let's reverse course.

The Electric Car That Almost Was

DURING THE FIRST decades of the twentieth century, electric, gas, and steam cars all vied for a lucrative spot in the future of transportation. The chug of steam, the grumble of gas, and the quiet ease of electric vehicles shared the road—with no discernible leader. The automobile race was in full gear, revving into motion with the best of intentions. Populations in cities were skyrocketing, and cars were a promise to city dwellers who desperately wanted to increase speed and efficiency while also eliminating traffic congestion, accidents, and pollution (read: horse shit).

"Every family in the United States in 1870 was directly or indirectly dependent on the horse," notes economist Robert Gordon.[68] As progress continued, populations in cities rose, as did the number of horses. Each stressed, skittish horse needed to eat 1.4 tons of oats and 2.4 tons of hay every year. Not surprisingly, pollution was reaching new heights, and people were drowning in the stuff. There was urine, manure, flies, carcasses, and all sorts of stenches and unpleasantries on the road. New York alone had 100,000 horses pooping out manure every single day. Rain turned the dung into a runny sludge that would swamp streets, while hot summer days parched the poop to dust, which the wind then whipped into a cloud that would choke pedestrians. A public health crisis loomed large. Horses had been a mode of transportation for thousands of years. Where were residents to turn for a cleaner city?

Automobiles were considered saviors in a shitty city. Only later did the irony hit with catastrophic force. Had the swap been a deal with the devil? A bargain with a wink? Yes and no. When steam, gas, and electric vehicles came onto the scene, they offered a breath of fresh air. Each had pros and cons, but at least none excreted goop from their back ends. For the car-crazy citizens of the day, the purchasing dilemma went something like this: To start a gas car, you needed elbow grease to turn the crank—and if it backfired you could break an arm. These vehicles were also dirty and loud. If you wanted a safer option, you could shop for a steam car, which for a few years was the fastest type of vehicle on the road, capable of reaching 127 miles per hour (this was the "Stanley Steamer," produced by twins Francis E. and Freelan O. Stanley). Steam had power, and, after decades of steam trains and boats, it was a familiar entity. But these cars were noisy, too, and they had issues with freezing. They required lots of patience; it could take up to forty-five minutes to heat the water in the tank and create the steam that would power the engine. Then there was the electric car with its rechargeable battery: clean, easy to use, and quiet. There were no cranks to be cranked or water to be heated up; you just turned a key and started driving. And yet, these cars had one fatal flaw: they were measly on mileage. They could go about thirty miles before they died and needed to be charged.

Thomas Edison—who would go on to invent the common electric light bulb (based on the innovations of inventors before him)—was a fan of the electric vehicle, advocating for its adoption. In 1884, he predicted that "the great secret of doing away with the intermediary furnaces, boilers, steam engines, and dynamos [to produce electric power] will be found, probably within ten years."[69] His optimism was too strong by more than a factor of ten. Edison's decade-long attempt to make a better battery ended with nickel and lithium. He never built a true lithium-ion battery; the element played only a minor role in his final design. Edison's battery was a success compared with previous models, but soon the automatic starter for gas-powered vehicles arrived on the scene and Edison's vision of an electric-car future was snuffed out. In the automotive race, gas-powered cars were the clear winner.

By 1929, most horses on the streets had been replaced with motor vehicles. Today, we fuel our cars by pumping the remains of ancient life like plankton into our tanks, drilling liquid graveyards in order to extract organic molecules (thus the name "fossil" fuel). When it comes to the repercussions of this habit, we too often turn a blind eye. Humanity, it turns out, has embraced the adage "I'll believe it when I see it," a misleading platitude at best. We may not see the levels of toxic carbon dioxide growing and gathering force in the air, but empirical, peer-reviewed testing tells us it is there.

This is also how we know that beneath our feet, Earth's core is a seething ball of iron the size of Mars with a temperature equal to the fiery skin of the sun. Above our heads, the stars are not just romantic night-lights but a series of explosions so colossal it will take billions of years for each dead star to use up all its fuel. Our existence on the surface of the Earth is dependent on the sweetest of sweet spots. Too much heat and our fragile skin burns. Too little and our tissue freezes black, slowly dying from the tips of our fingers and toes. What we have going for us is a well-balanced atmosphere that protects us from the radiation of the heavens and fills our lungs with life-giving oxygen. But this extraordinary envelope in which we live is swiftly filling with too much carbon dioxide.

Lithium-ion batteries may help change that. Currently, lithium is used in everything from medications to treat bipolar disorder to aircraft to batteries. In its purest form, lithium is silver white and soft enough to cut with a butter knife. It is one of the first three elements created in the minutes after the Big Bang. Despite this, there is less lithium in the universe than other elements on Earth. Our oceans contain 180 billion tons of the stuff, but it's so diluted, at about 0.2 parts per million, that there is no feasible way to extract it. Lithium is also highly reactive and flammable. If handled improperly, lithium-ion batteries can explode. Rest assured that we don't have to worry about the lithium in our bodies causing *us* to explode—the amount present is much too small—but to suppress the volatile nature of lithium in batteries, a separator must be placed between the lithium cathode and the anode.

A reliable lithium-ion battery was only made commercially available in 1991. Soon after, it became the go-to choice for personal electronics. Chances are your laptop, tablet, and mobile phone all use lithium batteries. Even the International Space Station uses them; its old nickel-hydrogen batteries were swapped out for lithium-ion batteries in 2017. This is no small task in space. It took eight years of research and fourteen space walks to replace the aging batteries as part of an electrical system upgrade. According to the Royal Swedish Academy of Sciences, these batteries are helping to shift "vehicles powered by fossil fuels to electrically-powered transportation." While we have yet to see the overall consequences of this development, "future breakthroughs will undoubtedly lead to further improvements in our lives, not only for our convenience, but also with respect to global and local environments and, ultimately, the sustainability of our entire planet."[70]

Lithium-ion batteries power our lifestyle, but a drawback is what it takes to make them: extremely high temperatures (lots of energy), environmental degradation during mining, and the huge amounts of water needed to extract lithium. This has people worried that electric cars are not as environmentally friendly as they appear. Most cars on the roads these days still use lead-acid batteries. Electric cars use lithium-ion batteries, but you'll likely find a lead-acid battery in there for the lights and audio system. These lead-acid batteries are up to 99 percent recyclable, if done properly. Lead is a potent neurotoxin, so any mismanagement or unregulated operations can have dire consequences. In 2015, Exide, previously one of the world's largest producers and recyclers of lead-acid batteries, shut down its plant in Los Angeles after it was discovered the company had been illegally releasing toxic lead dust around thousands of surrounding homes for decades, causing $650 million in damages to clean up and untold numbers of health issues.

At the moment, graphite is the anode material of choice for lithium-ion batteries, but scientists are looking to tap the potential of silicon. In principle, silicon can store up to ten times more energy than graphite, with each silicon atom binding to four lithium ions. The setback to this prospect is considerable, however: after only a few charging sessions, a silicon battery fractures and dies. To overcome this problem, Cui's team needed to find a way to protect the silicon nanoparticles from unwanted reactions.

Enter the pomegranates (modern French for "grenade") and their explosive introduction into the world of lithium-ion batteries. Cui's initial design in 2013 encased the silicon nanoparticles, or "seeds," in a highly conductive carbon shell. A year later, the design was modified to bunch the nanoparticles like pomegranate seeds clustered into a carbon rind. Pomegranates have their structure because they need to pack as many energy-rich, juicy seeds as possible into a tight enclosure. The hope for silicon "seeds" is to do the same and pack as many together as safely as possible, but silicon is difficult to use in batteries because it breaks easily and reacts poorly with other chemicals, forming gunk that kills the system. By protecting the silicon nanoparticles in a carbon rind—the way a pomegranate's seeds are sheltered inside the fruit's squishy pulp—the silicon "seeds" can swell and shrink without doing any damage.

Cui's work made it to the coveted cover of *Nature Nanotechnology* in 2014. His team's batteries retained 97 percent of their charge after one thousand charge cycles—well within the range for commercial operation. In his paper's abstract, he gave a hat tip to the humble pomegranate: "Our design is inspired by the structure of a pomegranate, where single silicon nanoparticles are encapsulated by a conductive carbon layer that leaves enough room for expansion and contraction following lithiation and delithiation."[71]

Cui's team has gone on to design other iterations, some that lean into the initial seed of inspiration and others away from it. Whether scientists will steer toward his innovation remains to be seen. Sometimes nature's riddles don't take us down the right road and sometimes they do. And sometimes we need to keep journeying to figure it out. Even if this battery is not the final design decades down the line, it contributed to the growth and evolution of ideas.

Growing Batteries

IN THE FALL of 2009, Dr. Angela Belcher handed Barack Obama a small card; on it was a picture of 118 squares, each with little notations inside. "Sir, I want to give you the periodic table in case you're ever in a bind and need to calculate molecular weight." The then president looked at the card and smiled. "Thank you," he said, "I'll take a look at it periodically."[72]

Obama had stopped at Belcher's lab as part of his visit to MIT. He was preparing to announce $2 billion in funding for advanced battery technology, and Belcher had recently showcased a super-tiny lithium-ion battery built using viruses—of all things—to assemble the positively and negatively charged ends. Employed on a large scale, this feat of engineering has the potential to reduce the toxicity of batteries, improve their lifetimes, and increase their charging rates. Obama's visit was an upgrade from when Belcher submitted her first grant proposal in 1999 and the reviewer told her she was "insane."

Belcher is a materials scientist and biological engineer, whose seed of inspiration was formed during her graduate years at the University of California at Santa Barbara, where she was studying how abalone (those snails that live in the sea) engineer their shells. Abalone construct strong, oval-shaped shells with a lustrous nacre on the inside that shimmers from silvery white to aquamarine and is pitted with a series of breathing holes. To make this exquisite material the snails have become masters at harvesting the resources they need from the ocean. Over millions of years, they have evolved key proteins that latch on to molecules like calcium in the ocean and deposit them on their shell in an ordered sequence. Maybe, Belcher thought, she could do something similar but on a much shorter timescale. Perhaps she could manipulate the DNA of viruses, which naturally latch on to organic matter like bacteria, to instead bind to inorganic materials like metal? If so, the viruses could help her achieve a coveted task in energy development: build minuscule battery parts using greener engineering. Conventional manufacturing requires high temperatures (up to 1,830°F/1,000°C) and toxic chemicals; viral batteries could be grown at room temperature using a water-based process. On the surface, viruses seem like a strange place to grab molecules and make stuff. When we think of viruses, we often think of something insidious, like the flu. But the truth is that viruses flourish everywhere, from the oceans to our guts, and Belcher is trying to harness these "heralds of death" to power our next generation of batteries.

Viruses straddle the line that separates what it means to be alive or dead. Most scientists put viruses in the "dead" category, even though they can adapt to their environment. Viruses are often likened to "androids" and "microscopic zombies" rather than living organisms.[73] Even though they carry information about themselves as either RNA or DNA, they have no way to mate and pass on this material to their offspring. To replicate, they must use their "tail" to stab the membrane of a host cell and inject their genetic material, piggybacking on the host's ability to make copies of itself. In her work, Belcher uses the "hijacking" ability of a virus called M13 bacteriophage, meaning "bacteria eater," because it thrives by infecting bacteria but is harmless to humans. The M13 virus looks like an extremely thin noodle with frills on each end. The virus is a nemesis to bacteria because the outside of its coat is covered in proteins that bind to proteins on the host's cells. Other viruses use a similar technique to make us sick, such as the flu virus that binds to cells in our respiratory tract. Once bound, the virus aggressively injects its DNA and takes control of the host cell's machinery, directing its host to do the one thing the virus cannot: reproduce. The hijacked cell releases a swarm of viruses that go on to seek other hosts to do the same. The result is an explosive proliferation of viruses that can, if it's like the flu virus, harm our health and well-being.

The ability of M13 to inject material into a bacterium makes them useful as genetic messenger drones. First, though, Belcher needed to find the best-performing viruses, and to do this, she needed to mutate a billion M13 viruses, each with somewhat varied genetics, in order to find those that bind to materials they don't normally bind to, like gold or cobalt oxide. She cherry-picked the viruses with the most promising interactions and did the process again and again until, generation after generation, she found the finest viruses dedicated to binding to battery materials. This was a decent step but still not good enough. Belcher then tweaked the virus's DNA to attract conductive materials like gold or cobalt oxide. This makes it sound easy, but let's break down what this looks like for cobalt oxide, a compound used in batteries. M13 has 2,700 copies of a specific protein on its coat called p8 (so named because it's encoded by gene 8 in the phage genome), and Belcher genetically added a gene sequence that gave the p8 proteins a negative charge, which means they can hold on to a positive charge. The virus's coat was now attracted to the positive charge of cobalt oxide or gold particles, and it covered itself in tiny pieces of metal (think of pollen stuck to the hairs of bees and you'll get the idea). To produce enough material to make a battery, she needed billions of virus copies. But since viruses don't replicate on their own, she exposed the virus to billions of bacteria and the virus swiftly replicated inside its bacterial hosts. Later, the bacteria are removed, leaving just pure concentrated virus. When linked together, these virus particles make battery materials like beautiful cobalt oxide nanowires—the key here is the ability to create incredibly lightweight, ultra-tiny batteries. Each little wire is about 80 nanometers in length, or the width of a red blood cell. Belcher can then repeat this process to build another part of the battery, this time exposing the virus to something like iron phosphate covered in carbon nanotubes, or manganese oxide, a material used in a lithium-air battery's cathode. The upshot? Belcher is essentially employing viral androids to do the assembly-line work of making materials for tiny batteries.

For the finishing touches, an assistant uses a pair of tweezers to assemble the battery components in a small coin-sized case like the kind that slots into watches. The viral battery is a layered creation, beginning with a disk of lithium foil as the anode, a few drops of electrolyte solution, a plastic separator, a few drops again of electrolyte solution, and a virus-based cathode, the ends capped shut with a flat, circular case.

So far, virus-powered batteries work as well as or better than traditional electrodes. And to think: Belcher has room to play with the genetic parcels in her process, coaxing the viruses (so far) to work with more than 150 materials. With the Koch Institute for Integrative Cancer Research, she is also looking to use genetically engineered viruses to detect ovarian cancer cells. Detecting this cancer during its early stages is challenging, and failure to do so contributes to a patient's low survival rate. By using viruses that carry a material that fluoresces, or lights up, and attaches to cancer cells, Belcher hopes small tumors can be detectable with imaging scans and help doctors to spot the cancer early. It's possible, then, that her "heralds of death" may turn into "heralds of life."

Body Fat–Inspired Batteries

LET'S EXPLORE ANOTHER "herald of life," one that, like viruses, has a bad reputation: fat. It's a loaded word these days, but fat is crucial for most of Earth's creatures. The bulk of us—humans included—store fat in various places around our bodies. Few animals, apart from outliers like camels, stockpile all their fat reserves in one place because it's usually not the most functional approach. Fat distributed around the body is more "fuel efficient," keeps us balanced, protects us from impact, and allows for a variety of animal shapes, since there doesn't need to be one large "pocket" for all the fat. Yet when we build robots, that's exactly what we do—one big energy battery sits at the robot's core.

Recently, a team from the University of Michigan looked at that one big battery and asked, *Why?* The limitations sometimes far exceed the benefits. A robot's battery takes up as much as a fifth of its internal cavity and requires a lot of costly energy transfer. Can a robot have a distributed energy system, like how an animal stores fat? That's the question Nicholas Kotov's lab team wants to answer.

A fair-haired man with a soft, stuttering voice, Kotov was born and raised in Moscow to a chemist and a physicist. His interest in science began at an early age, at one point leaving a scar on his hand when an attempt to stuff matches into a tube to make fireworks resulted in the fir-sap seal melting. As a child, he says he felt immense pressure to be like everyone else, but his stutter was a constant reminder of why that wouldn't be the case. "I was not part of the herd," he writes on his faculty page. "Only much later did I realize how useful that actually was."[74]

Kotov is now an Irving Langmuir Professor of Chemical Sciences and Engineering at the University of Michigan. His lab has created materials for thinner lithium-ion batteries, safer paints, and cancer diagnostic tools. At the heart of his work are nanomaterials, where a single nanoparticle is 100,000 times smaller than the width of a human hair.

Self-assembly of nanoparticles is a new field that explores the interactions of these tiny players in their microcosmic bubble of a world. Kotov believes that an understanding of nanoparticles and the elegance of nature's laws is the key to pushing the boundaries of energy and engineering. And he's using his super-small expertise to disrupt the paradigm of today's batteries and create distributed energy storage for robots. His batteries look more like a shell around the robot; in this way, the batteries can meet the form and function of the robot, rather than requiring scientists to engineer a robot around a single bulky battery. He may even be able to shrink these batteries to the microscale one day.

His team has proven that fat-inspired batteries can work by putting them in, of all things, a crawling robo-maggot. They call their mechanical bug a caterpillar, but almost everybody sees a maggot (thank goodness Kotov didn't go into marketing). For good measure, and to showcase the batteries with a variety of forms, they also created a robo-scorpion, a robo-spider, and a robo-ant. In his experiments, the "biomorphic" batteries—or those that resemble living beings—generated up to seventy-two times more energy compared with a single lithium-ion battery of the same volume. The rechargeable zinc battery is also made with relatively cheap and nontoxic materials. The stumbling block is that they can do only about a hundred cycles before they deteriorate, whereas lithium-ion batteries can do more than five times as many.

Distributed batteries also provide a robot with a cushion similar to fat. "Fat serves thermal functions and structural functions. If you were to fall on just bone, the damage is big. If you fall on a cushion of fat, the damage is less," says

Kotov. "Many organisms learned this, and the same idea is applicable to our technology." A distributed design could also be useful to extend the range of cars, says Kotov. "Upward of 20 percent of the weight of Tesla, at least of the older models of Tesla, is the battery. It's a good case for where optimization is needed."[75]

The batteries free up both space and weight, while also increasing energy capacity. Kotov has just begun talks with commercial partners and hopes his lab's work can hit the market in the next few years. He sees the possibility of his research being used for small robots that self-assemble into larger devices once they reach their destination, as well as delivery drones, takeout bots, robotic nurses, and warehouse robots.

Dr. Fumiya Iida has a similar vision. Iida is a researcher in the Bio-Inspired Robotics Lab at the University of Cambridge, U.K., whose goal is to engineer robotics that use energy more efficiently. To perform the same action as an animal, robots require between ten and one hundred times more energy. As he told me, "Robots can do incredible things: they can do fast motion, precise motion, and so on, but at the expense of efficiency. If you look at biological systems, our evolutionary process has had to really consider energy efficiency because energy is one of life's fundamental bottlenecks."[76]

Energy efficiency is one of nature's main design principles, whereas historically it hasn't been as crucial in robotics, says Iida. If you look at the engineering world, you'll see that locomotion is mostly done with wheels, whereas animals walk, flap, and swim. A robotic leg is difficult to design and, when we've done it, the result has been energy inefficient.

Although wheels excel on a flat, smooth road, wild country is often pockmarked with ditches, mud pits, ruts, hills, rocks, and more. A walking or crawling robot can move in varied terrains and not get stuck. But today's "robots are burning lots of energy just by standing, just by walking slowly, whereas in nature we somehow use almost no energy for walking. And walking for animals is almost the same energy efficiency as locomotion on wheels," says Iida. "So how this is possible is a big question for me."

One answer is that our anatomy usually serves multiple purposes. Muscles don't just generate force like a motor; they also store energy like a spring, act as shock absorbers, generate heat, protect our bodies, and have sensors inside. Bones too are multifunctional, providing structural support for our bodies as well as storing calcium and producing blood cells. Another fundamental difference between humans and machines is that some 90 percent of the human body is made of soft substances, whereas machines are 99 percent rigid. They're made out of metals, aluminum, and the like.

"The biggest challenge I'm trying to address is the complexity issue," says Iida. "How can we make a complex body of robots?" It's a question he's still exploring. The natural world, including all of its bacteria and viruses, is a Morse code of amino acids that build into billions of forms. It is a system of parts we are all novices at understanding, and it takes time and patience to unwrap these discoveries like a sweet fruit.

The search for ever more powerful, safe, and long-lasting batteries with new chemistries and designs continues. BloombergNEF predicts that more than half of all passenger vehicles sold by 2040 will be electric, meaning 31 percent of cars on the road will be electric vehicles.[77]

As Sheikh Yamani, who served as Saudi Arabia's oil minister three decades ago, said, "the Stone Age came to an end not because we had a lack of stones, and the Oil Age will come to an end not because we have a lack of oil."[78] There is irony in him saying those words, but perhaps a kernel of truth, too. Maybe it is both necessity and the emergence of technologies that surpass oil's power that will usher in a new age of transportation.

More than two hundred years after Alessandro Volta invented the "pile," we are at the nexus of another revolution.

9

SKELETONS IN THE CLOSET

Bones Inspire Lightweight Aircraft and Architecture

"Adapt or perish, now as ever,
is nature's inexorable imperative."
H. G. WELLS, author

OF ALL THE rooms in a natural history museum, it is the back archives I love the best. Open a drawer at random and bat skulls with empty eye sockets stare back at you from plastic tubes, identification tags twined around the caps. Open another drawer and wing bones are spread out like gaunt white fingers; another and huge butterflies are pinned like iridescent blue-green bows. Across the corridor is a boneless, cartoonish tiger, as if the taxidermist had looked at the Broadway production of *The Lion King* for inspiration. Few places inspire such a curiosity and a grievousness in equal measure: a wonder at the variety of shapes that evolution can sculpt and a heaviness in knowing that many of these creatures met their fate at the hands of humans.

I remember, as a child, picking up a deer vertebra in a friend's backyard and turning the bone in my hands.

I was awed and saddened, two emotions I didn't know could coexist until that moment. *Why are there so many holes in their bones?* I wondered. *How do these hard parts form inside a mushy body?* Similar questions come flooding back to me on a visit to the archives at the California Academy of Sciences. I'm here to examine their black-market collection of trafficked creatures and to introduce myself to their in-house scientists. In these grim aisles, I can't help but notice the versatility of bone. There are the branched antler bones of moose that battle each other for a mate, and the toe bones of a bat that grip a tree as it snoozes upside down. Even the finger bones I use to pry open the museum drawers are works of evolutionary art, providing the dexterity to use tools that helped change humanity forever.

Despite common belief, the planet's boneless creatures far outnumber the bony; 96 percent of animals have either soft bodies (like jellyfish) or exoskeletons made of chitin (like insects). For the remaining 4 percent, the vertebrate kingdom, bones are the hidden architecture. This was not the case 550 million years ago, when there was no such thing as a skeleton. All the animals were jawless, spineless, squishy ocean dwellers. A cascade of events began to change this 1.5 billion years ago, when the violent movement of tectonic plates washed large amounts of calcium carbonate minerals into Earth's oceans, allowing marine creatures to test out a new molecular tool kit over millions of years, eventually stumbling upon a combination of ingredients that gave them a bone-like substance. The mineral's arrival allowed animals to grow hard parts on the outside of their bodies. At first, these bone deposits may have been a reaction to the sharp uptick in minerals swirling about, helping the creatures to rid themselves of too much calcium, which is toxic at high levels. The deposits were toxic waste dumps, of sorts. Soon, however, they proved fruitful. Bony armor flourished as part of an evolutionary arms race, protecting soft bodies from predators.

About 10 million years after the introduction of calcium carbonate into Earth's oceans, a flurry of multicellular life entered the geological record in what's known as the Cambrian explosion. The trick of bone proved a successful one. Skeletons evolved more than three dozen times, becoming the "must-have" accessory of the era. And yet, the development of bone had its drawbacks: the newly minted parts did not allow for much movement. The second shift in evolution came when these hard parts sank *into* the animal's body, becoming a support structure for tissue while also serving as protection for soft organs like the brain. For prehistoric fish, skeletons were an internal repository of minerals, the bone cells becoming something like skeletal batteries to allow fish to travel greater distances in the ocean.

For our part, the written record reveals our confusion and ignorance of bone. We imagined bone as something white, dry, and lifeless, dead material dug out of the dirt or bleached and displayed in glass cases. Galen, a physician and philosopher in the Roman Empire, proposed the startling idea that bone, because of its pale color, is made from sperm. In the eleventh century, Persian polymath Avicenna said bone was primarily cold and dry, like the earth itself. But overall, medieval and early Renaissance anatomists didn't have much to say about the skeleton. It seemed too simple and self-evident, the domain of less learned practitioners who dealt with broken bones. That would soon change. In 1506, Leonardo da Vinci found himself dissatisfied with humanity's knowledge of bones; he began anatomical explorations, starting with his dissection of a one-hundred-year-old man. For the next six years, he dissected more than thirty human corpses and illustrated almost every bone in the body. An incredible anatomist, he filled page after notebook page with precise drawings and measurements. Although none of his work was published, by the end of the sixteenth century the skeleton had become the classic image of human anatomy, and, when draped in a grim reaper cloak, death.

What anatomists in the early Renaissance couldn't know is that the skeleton inside us is neither sperm nor earth nor death. Our bones are alive. Sunk beneath our skin, in seclusion from the outside world, bones are vibrant tissue vital to our survival, and nowhere is this more clear than when bone growth goes awry. A patient with brittle bone disease, for example, can sustain fractures simply by sneezing. Patients with Münchmeyer disease, on the other hand, produce too much bone, and trauma to their body results in extra bone struts that can lock their arms and legs in place. They truly become trapped by the bone in their bodies. Then there is skull shaping, a cultural practice that has been around for some 5,500 years and was seen on every inhabited continent. There were the flattened skulls from certain Native American tribes, where an infant's head is secured to a cradleboard. There were the elongated skulls preferred by the Huns, Maya, and Pacific Islanders, who bound their children's heads as they grew, squashing their skulls. It has been seen in France, Russia, Scandinavia, and others.

Slowly but surely, the life of bone crept into our collective consciousness. In 1939, there was even a case where Dutch anatomist E. J. Slijper found a goat with two nubs for front legs that had improvised its own style of getting around, hopping somewhat like a kangaroo. After the goat died, the anatomist dissected the animal and saw that its hip and leg bones were thicker than usual, hinting at the possibility that the goat's skeleton had been growing bone where it would be useful for hopping.

Decades later, we learned how this is possible. Bones contain a crew of cells that are constantly busy constructing, maintaining, and remodeling our skeleton. It takes a village to scaffold the bone in our bodies, to create an arrangement of levers, columns, and hinges to perform weight-bearing movement. There are two main types of bone tissue: a hard, compact bone on the outside (cortical) and a spongy bone (cancellous) that looks a bit like Swiss cheese on the inside. Both are found in most bones, but the spongy variety makes up about 20 percent of the human skeleton and is mostly located at the ends of bones and joints, providing structural support and flexibility. Compact bone makes up the rest as a shell around spongy bone and is the primary component of the long bones in our arms and legs, where greater strength and rigidity are needed.

You can tell a lot about a person by just their bones. Take me, for example: If you were to examine the biography of my bones, you would know my height, my gender, and roughly my age. You would see a healed fracture in two places on my right arm, and trauma to the arch of my left foot. These breaks happened young and healed themselves with the aid of a cast, thanks to one human physician and many cellular physicians hurrying to the scene to mend the gap. By now, the cells that were crucial to my recovery have perished, their lives fleeting in the grand scheme of things. Not one inch of my skeleton is the same as it was ten years ago. The human body replaces its bones every decade. That's about 42 billion osteocytes—star-shaped bone cells—that are new in the skeleton, and this doesn't even include the three other types of bone cells: osteoblasts, osteoclasts, and lining cells. It's truly a party—or, more accurately, a work conference—in the body. If you were to grab a chain saw and cut through bone, you'd see why. It's a masterpiece in efficiency, or, as materials scientists would say, optimization.

The general theory for how our bones remodel is called Wolff's Law; it refers to the way in which the body responds to stimuli such as running or skipping rope by putting down more bone tissue. If there is less strain on a bone from lack of use, the osteoclasts break down the tissue, releasing calcium from the bone to the bloodstream. This means our skeletons adapt to the burdens we place—or don't place—on them. For once, stress is actually good for your health!

Modern life has certainly left an indelible mark on our bones. Anthropologists refer to human skeletons as gracile, meaning we have less bone mass for our size than our ancestors or closely related wild apes. Our skeletons have not caught up with our cultural and technological milieu; the transition to a sedentary lifestyle has chiseled away at us and made us more susceptible to fractures and bone diseases. A 2009 study in Germany found that the bones of children were lighter and more fragile than children's bones just a decade before. When the team controlled for other factors, they found a strong correlation between the amount of exercise a child got and the heft of their bones. The current composition of our bones may owe more to our choices than to evolution.

A more dramatic example is a spike-like feature found on the lower back of the skull in some people. As we hunch over our computers and phones, we crane our necks forward. Our ten-pound head is no longer aligned with our spines, and the posture puts extra pressure on the neck muscles attached to the skull. The body responds in turn by adding more bone cells to cope with the added stress. It's a powerful realization: our actions can change the constitution and shape of our skeleton, to a degree.

The Bare Bones

WHEN I FIRST visited the archives at the California Academy of Sciences, my purpose was purely investigative. I was there to introduce myself to their scientists so they could contact me if an illegally traded specimen was seized by police and sent to their department. At the time I was doing research into wildlife crimes, but a year later, bones cropped up again in my life, this time as I was researching material for this book. A silver-haired man with dark glasses named Jeff Brennan shared with me how our bones are inspiring engineers and designers to make lighter inventions. The story begins in the early 1990s, when a younger, darker-haired Brennan was a student all lined up to go to medical school. His trajectory was perfectly optimized: he had taken all his required courses, had dissected cadavers for his anatomy class, and was working on artificial limbs and prosthetics for amputees. Then there came a point where he realized this wasn't his dream. He transferred to the University of Michigan to do a PhD in orthopedic biomechanics, investigating how our bones grow and remodel.

The idea that you could somehow mimic the efficiency formula of bones using mathematics transfixed Brennan, and he began collaborating with his professor, Scott Hollister, and Dr. Noboru Kikuchi, who was one of the founding fathers of topology optimization, a method to mathematically optimize the layout of a material within certain design parameters. In the late eighties, Kikuchi was trying to model how the human body grows bone, and whether there was an optimal solution to its layout. Does the body follow rules in how it responds to stresses and strains? Could he translate the growth patterns of bones into a mathematical formula? If researchers could model how bone grows, perhaps they could use that knowledge to stimulate healthier bone growth in later life. This could help those with, for example, osteoporosis, a disease in which bones become brittle, and a common condition in women after menopause.

The hope was to develop a therapeutic tool, but the kernel of an idea took on a life of its own. Attempts to enter the medical innovation market are notoriously slow due to the many regulatory hurdles that must be cleared in order to establish the safety of new devices. Brennan didn't want to wait that long. He wondered whether we could use this mathematical formula for something other than the medical world. After graduate school, he went on the hunt for a job where a bone-inspired formula could be put to good use. One of the employers he came across was Altair, a young company that provided software and cloud solutions. Altair was the riskiest of the three companies that Brennan was looking into because they were relatively young, and the field of topology optimization was nascent, but they shared a common vision. Brennan and Jim Brancheau, chief technology officer, "spoke about the possibility that in the future, you could use this type of math to grow ideal shapes of structures for everything from automotive brackets to you name it," Brennan told me.[79] Brennan joined the company, and it wasn't long before he sent CEO Jim Scapa a paper by Dr. Kikuchi about bone optimization and asked, "What do you think about this?"

Scapa replied with a sticky note on Brennan's desk: "Do you think we can commercialize it?" Brennan wrote "Yes" and returned the note. With Altair's blessing, Jeff Brennan assembled a team to make a software product based on the mathematical formula of bone formation. They named it OptiStruct, a fusion of "optimal" and "structure." Brennan admits he isn't the "genius technologist" behind the idea, just an adoptive parent, and the way he waxes poetic about the software certainly makes you believe him. Apart from his kids, the software is his self-confessed favorite topic in the world.

OptiStruct helps customers optimize the weight and strength of a design—such as an aircraft wing, car part, or bike frame—in response to stress. The software calculates where to add or subtract material for optimal performance. Their algorithm tests a range of options and lands on the best solution between the amount of material, strength, and ease of manufacturing. The result is a design that uses fewer materials and typically takes less time to produce, because it reduces the number of product iterations down the line. The software treats the volume of an object as if it is already a porous medium with holes. Where there is more need for support in the design, it reinforces that area and makes it stronger. Where less support is needed, the material becomes weaker by expanding a hole.

If you think about the three Rs (reduce, reuse, recycle), OptiStruct is all about the first: reduce. If you save, say, 20 percent of a plane's or train's weight ahead of construction, you never have to mine that extra material, transport it, process it, carry it around on a vehicle for twenty to forty years (saving on fuel costs), and eventually dispose of it. It never existed—and there is something beautiful in that. The software encourages a minimalism of sorts, a shedding of the excess bulk from our structures. The technology forces engineers to step back and say, "Okay, what do I really want? What are my constraints? What are my variables? Is it overdesigned? Where can we punch holes in this thing?"

For example, with a few taps on my computer keyboard, an arched vault for a primary school in rural China forms on a simulation screen and then swiftly dissolves into the lithe skeleton of its original form. The school is tucked into the remote mountains of Lushan, where steep peaks capped with Buddhist and Taoist temples rise above a sea of clouds for nearly two hundred days of the year. The design for the school is composed of veinlike lattices, a mixture of straight lines and arches that remove all excess bulk. It is a hollowed-out version of its original form. The arrangement is either eccentric or elegant, depending on your aesthetics, but nothing is wasted. The arches are strong and thick where needed, just like our bones, and light where it's not. When finished, the school will serve 120 pupils from twelve local villages. Few will know human bones inspired the willowy look. One reason is because design often begins with adding in mind (just think of the name "additive engineering"); the consideration of negative space is more an artist's modus operandi. When painters put brush to canvas, they consider the area *around* a subject. The negative space is just as crucial as the positive space. And yet, few other professions dip into this mindset, even for a thought experiment. Our bodies, however, do operate this way, purging the excess of our bones, keeping only what's necessary for a strong, light, and flexible frame. Raw materials in nature are not a guarantee, so it's all about doing more with less. A spider weaves a web that is resilient to the whap of a fly, but it doesn't produce more silk than it needs. "In her inventions," said Leonardo da Vinci of nature, "nothing is lacking, and nothing is superfluous."[80]

Less Is More

OPTISTRUCT WASN'T AN easy sell early on. Brennan spent much of his early time at Altair schlepping a computer around, trying to persuade engineers to believe in his idea. Brennan sees "lightweighting" as a way of thinking, a design mindset. Ford commercials, for example, don't talk about how light their new vehicle is compared with previous models. "They talk about industrial-grade aluminum. Tough guy stuff, right? But the truth is that their truck was like five hundred pounds lighter than the previous. They didn't market it for that, 'cause that's not what sells trucks. It's green stuff," says Brennan. But the company still decided to work with Altair.

Ford's decision went against mainstream models. At the time it was not "less is more" but "more is more." And, to be fair, not everything that is mathematically optimal is cost-effective or suitable for manufacturing. The software can advise the best architecture and geometry with the least amount of materials, but that doesn't necessarily mean the shapes can be molded using traditional methods. It's a formidable stumbling block. Altair needed to find a way to make the software operate within the limits of human potential. Nature prefers to design in spirals, bends, and curves; humans are linear in their construction. 3D printing could help with this, but that technology only emerged on the scene in the 1980s, and it was expensive, with few materials that could be harnessed. To take these limitations into account, Altair added manufacturing constraints into its software so users could specify the methods and materials they wanted to use. For example, if a designer is using ply, which is folded or laminated material, they can include in their design constraints how thick their ply material is, its orientation, and its stacking sequence.

Altair also partnered with Airbus, a leading aviation company, to determine the best layout for the ribs in the leading-edge wing of the world's largest passenger aircraft—the new superjumbo A380. Since even a few pounds in the air can impact fuel efficiency and emissions, the team at Airbus wanted to see if they could make the A380 more efficient from the blueprint stage of design. OptiStruct provided a template and Airbus engineers interpreted these results with their aesthetic vision in mind. Together, they saved over 1,100 pounds of weight per aircraft.

"That put us on the map in aerospace," says Brennan. The Boeing Company soon purchased the technology for the ramp of its CH-47 Chinook rotorcraft. The resultant open truss structure was stiffer than the original; it also saved 17 percent in weight and cut down on development time in the brainstorming room. For Volkswagen, Altair designed an engine mount bracket that weighed 23 percent less than its predecessor. Companies such as Lockheed Martin, Harley-Davidson, Procter & Gamble, Nokia, Adidas, Kohler, and Fisher-Price have also used this nature-inspired technology, saving billions of pounds of material each year and preventing even more carbon dioxide from entering the atmosphere. In 2009 alone, Altair estimated its software saved 1.3 billion pounds of material.

More recently, the software was used to help design optical satellite components for a company that develops low-orbiting and geostationary satellites for Earth observation, human spaceflight, exploration, and communication. Computer-simulation software helped predict load conditions during the launch phase of the satellite and checked its strength and stability. The potential to transition the design from the drawing table to real-world use is in large part due to the convergence of 3D printing and topology optimization. "They're made for each other," says Brennan. If the software creates a complicated structure and you have to print 300,000 of them for full production, it isn't going to happen—at least not now. But if you're sending one of these complex designs up into space and weight is critical, well, then that's a different story.

The software is a tool in the toolbox, something that can be chosen for specific projects. In an ideal world, greener materials and more ecological creations would be the norm. We're not there yet, but innovations like OptiStruct have the potential to get us closer.

King Kong's Broken Bones

IN CREATURES AND construction, there are limits to the size of things. You never see a gazelle supersized to the height of a skyscraper or a hummingbird swollen to the dimensions of an aircraft. Why? The first person to pose this question was the brilliant Galileo Galilei, who, in his final written work, *Dialogues Concerning Two New Sciences*, pondered why animals can't be any arbitrary size, like a horse-sized flea or a flea-sized horse; or why a flea can jump many times its weight but a human cannot. This deceptively simple question has profound consequences for design and construction. We'll circle back to human engineering soon, but let's first explore the limits to size in nature.

Size matters to almost everything in nature because the physical rules are different for each order of magnitude. This is why you can toss an ant from the Empire State Building and it will land unharmed but the same cannot be said for a horse or a human. The concept was popularized by J. B. S. Haldane's essay "On Being the Right Size." The smaller you are, the less concerned you are with big spills. Why is this the case? Let's examine a creature the size of a sesame seed and swell its height tenfold. A common mistake is to assume the rest of its features grow ten times as well. In fact, its skin will increase one hundred times and its insides a thousand times. This is known as the square-cube law, where an animal's insides grow more than their outsides.

If we took fictional characters like King Kong and placed them under the physical laws of the real world, the gorilla wouldn't be frightening at all; in fact, King Kong wouldn't even move—his bones too weak to support his mass and snapping like twigs. Another improbability is the giant in "Jack and the Beanstalk," who would have suffered painfully from leg bone fractures. At this point, you may be thinking, *What about dinosaurs?* The largest dinosaur that ever existed was *Argentinosaurus huinculensis*, a 130-foot (40-meter) sauropod with a tremendously long neck and hauling around some 50 tons of weight during the Late Cretaceous period in what is now Argentina. Austere and yet eye-catchingly eccentric, its proportions look comically confused, as if a child scribbled a dino from their imagination, concocting a creature with a stubby body that's overshadowed by an impossibly long neck. Shouldn't the dinosaur's legs shatter under its own weight? Considering the largest African savanna elephant on record was 24 feet (7 meters) in length and a hulking 12 tons, the sauropod was actually pretty light for its size. Size limits in nature are only true if no innovation happens; that is, if nothing changes in the creature or the environment. Evolution found a way to sidestep the scaling issue in dinosaurs by developing lightweight bones with lots of space inside, especially the bones in their neck, some of which, like bird bones, stored air and were extensions of air sacs in their lungs. These pneumatized bones, as they are known, are much lighter than your average elephant's bones. Then there is the blue whale, which is heavier than any dinosaur, coming in at a whopping 150 tons. But comparisons between dinosaurs and whales are inherently flawed because the physical laws that affect the two creatures are different. The buoyancy of salty water frees the whales from the constraints of gravity that terrestrial creatures must endure.

If we return to the square-cube law, the physics of scaling also holds true for machines, buildings, and their supporting structures based on their materials. As we scale up a building or a creature, its weight increases in proportion to its volume. In *Two New Sciences*, Galileo gave the following example: If two ships are built, one large and one small, with identical scaling proportions, the large vessel must diverge from their similarities to provide more scaffolding and support; otherwise it will collapse under its own weight. Another way to imagine this is to think of a chandelier suspended from a ceiling by a thin rope, just strong enough to support its weight. A man walks by, sees the chandelier, and wants to copy its vintage design for his dining room, but to do so he must double the size of the chandelier to fit his larger home. The man would like the same chandelier scaled twice as large in every dimension, increasing its volume eightfold. He might then conclude that he needed to also double the dimensions of the rope to hold up his new chandelier. But the weight-bearing capacity of the rope depends not on its volume but on its cross-sectional area, or the number of fibers used to hold the chandelier. In doubling the dimensions of the rope, this cross-sectional area increases only fourfold, while the volume of the chandelier (and hence its weight) has increased eightfold. For the chandelier not to crash to the ground, the rope must be considerably fatter.[81]

Similarly, the strength of a pillar holding up a building is proportional to its cross-sectional area. A twice-enlarged pillar can hold four times the weight, but if the building it's supporting is similarly doubled in all dimensions, the building's weight will increase eight times, making the column crumble. Scaling plays a critical role in the design of bridges, airplanes, ships, and more where going from a small model to a large one takes expertise, especially to do so in an efficient, cost-effective manner.

Without scaling, we cannot understand the laws of nature. As Galileo says in his discourse, after chatting about the strength of ropes and beams, "you can plainly see the impossibility of increasing the size of structures to vast dimensions either in art or in nature; likewise the impossibility of building ships, palaces, or temples of enormous size in such a way that their oars, yards, beams, iron-bolts, and, in short, all their other parts will hold together; nor can nature produce trees of extraordinary size because the branches would break down under their own weight."[82]

From the tiniest organism to the loftiest towers, scaling is crucial in everything. When we try to achieve architectural feats that have never existed in nature, we do this by adapting to constraints to avoid a catastrophic collapse.

Jeff Brennan is not alone in finding inspiration in bone. Before him, in 1866, Swiss engineer Karl Cullman copied the structure of bone for the design of a crane. Nearly thirty years later, architect Gustave Eiffel crafted the Eiffel Tower based on a human thigh bone. Today, the monument is a symbol of Parisian culture, but at the time it was roundly criticized. Author Guy de Maupassant called it a "giant ungainly skeleton upon a base that looks built to carry a colossal monument of Cyclops, but which just peters out into a ridiculous thin shape like a factory chimney." Novelist and art critic Joris-Karl Huysmans said it was "a half-built factory pipe, a carcass waiting to be fleshed out with freestone or brick, a funnel-shaped grill, a hole-riddled suppository." A French poet was on the mark about its inspiration but didn't approve of the aesthetics, calling the tower a "belfry skeleton."[83]

In the twenty-first century, we've come to see the Eiffel Tower as beautiful, and expanded our fascination with bones beyond *Homo sapiens*. Why, some scientists wonder, do black bears not hobble out of their dens after eight months of hibernation, their bones as frail as breadsticks? During a bear's hibernation, the thump-thump of their heart slows, and they no longer urinate, defecate, or eat. Their bones *should* weaken from lack of exercise, but they don't. If they were anything like humans, they would suffer from diabetes and osteoporosis, their bones fracturing like chalk once they stood up and rubbed the sleep from their eyes. If a human behaved like a bear, she would experience bone loss, ketosis, hyperglycemia, and muscle protein atrophy—a whole bunch of physiological states that tell you your body isn't doing well.

The nearest humans get to this hibernation state is in zero gravity aboard the International Space Station. To keep their bones and muscles strong, astronauts must exercise for two hours each day, strapped to treadmills and bikes, and even doing so, they can still lose 1 to 2 percent of their bone mass per month after six months in space (the longest stay in space is nearly a year). During a six-month mission, some individuals have lost as much as 20 percent of the mass in their leg bones. So how do bears keep their skeletons strong as they snooze?

Scientists at Augusta University in Georgia collected blood and bone samples from thirteen female black bears before, during, and after hibernation for three years; they also measured their enzyme and hormone levels. They wanted to see what, if anything, changed. Back in the laboratory, the team spun the bears' blood to isolate the serum—a protein-rich liquid—to determine what was happening inside their bodies. The verdict? The bears stop calcium from leaving their bones during hibernation using a series of well-orchestrated mechanisms. Rather than growing new bone to prevent their skeletons from weakening, black bears press pause on bone loss. During their epic snooze, there is a fifteenfold increase in a protein called CART, which slows the process by which calcium is lost from their bones. At the same time, the number of bone-building osteoblasts plummets, as do two other enzymes that contribute to bone generation. This ensures the osteoblasts don't go haywire and build too much bone, throwing the balance of osteoblasts and osteoclasts off-kilter, as can happen in conditions such as type 2 diabetes.

In some diabetic patients with fragile bones, it's as if their body's bone optimization has failed. Lamya Karim, a professor at the University of Massachusetts, works with cadaver bones from donor banks as well as elderly patients undergoing hip replacement surgery. Karim collects the scrapped bones and studies them at her Bone Biomechanics Lab. She found that the bones from patients with type 2 diabetes are dense, but also more prone to fractures than those without diabetes. You'd think that dense bones would be stronger bones, but that's not the case here. These particular bones have developed too many cross-links, or thin bridges of bone. And those cross-links prevent the bone from being flexible, making it too rigid and brittle. The more *unnecessary* cross-links you have, the weaker the bone is. The osteoclasts, whose work is to break down old bones, have become sluggish in patients with diabetes. Their bone's ability to optimize has failed, and it has "overdesigned" its function, a predicament often seen in our buildings and inventions. What works best is, often, the "bear" necessities.

Can We Print Bone?

HAVING MADE INROADS with their bone-inspired software, Altair is now looking to design medical implants. The idea would bring the company full circle: a bone-inspired design to regrow bone. They now see the trifecta of 3D printing, stem cell research, and topology optimization as offering a promising future for 3D-printed body parts. So far, Altair has designed a prototype hip prosthesis. Titanium is the material of choice for implants at the moment because it is durable, and yet it's also 6.5 times stiffer than the dense bone in our femur. This means the implant doesn't obey Wolff's Law or topology optimization. To compensate, the bone changes around the titanium implant, shielding it from stress and causing bone loss, which can lead to fractures and dislocation. By designing a titanium implant that mimics the optimized structure of bone, the Altair team hopes the implant will last a lifetime and provide patients with a better quality of life. When the team simulated the stress of a generic implant compared with an optimized one, the latter showed stress concentrations more similar to healthy bone. The team is also excited about the possibility of adding personalized parameters. For example, if a two-hundred-pound male wants to water-ski and cycle with his kids, the implant could be optimized for his lifestyle.

The future may also see this technology merge with the work being done by companies such as EpiBone, a start-up looking to create bone from stem cells for transplantation, in which the bone would be grown on a scaffold that is optimized like the patient's own. We have cells in our bodies that grow bone, so why not try to collaborate with those cells in the laboratory to grow new tissue? It's an exciting prospect, but the regulatory pathways for these innovations are slow; it will likely be years before anything reaches the market.

Despite the formidable road ahead, it's inspiring to think about the millions of bones molded by evolution into a tapestry of shapes and the inspirations that await discovery. When we perish, our bones will likely linger, dead but present, far longer than our flesh and blood—a truly incredible feat of evolution.

10

A MONSTER IS BORN

Reptile Spit Inspires a Type 2 Diabetes Medication

"I believe that the more clearly we can focus our attention on the wonders and realities of the universe about us, the less taste we shall have for destruction."
RACHEL CARSON, biologist and author

A FAT SIAMESE CAT sits on a scuffed wood porch, silently observing a tanned man in gym pants and khaki boots as he's handcuffed outside his home at a dead-end gravel road in the rural foothills of North Carolina. The man had heard a knock, cracked open his door, and saw authorities on his porch. With a frown, he confessed: "I guess you are here for the opium."[84]

The authorities were not, in fact, there about the opium. They'd stopped by for an unrelated complaint, but the man's reaction piqued their interest. The deputies didn't have a warrant. But the man didn't ask to see one. He led authorities to his acre of land, where his garden was brimming with waist-high stems tipped with bulging seed pods the size of chicken eggs. Pops of mauve colored his garden. All told, the authorities confiscated some $500 million worth of opium poppies. The plants are harmless—unless you know what to do with them. With a razor, you can slice the seed bulbs and wait a day for drops of milky sap to ooze out. Then you scrape off the "white gum," as it's called, and process it with water and solvents to get a morphine solution.

This real-life bust in North Carolina draws attention to something we often overlook as we sift through our medicine cabinets: some of our most powerful medications come from the natural world. We've compared herbal remedies to witchcraft so often that we forget how many of our prescriptions are, in fact, inspired by the trees, flowers, and scrubby bushes of our planet. Of course, the man in question was likely not peddling morphine so much as heroin (which takes more chemical additives to concoct), but morphine doesn't just come from drug dealers; it's also piped into hospital intravenous tubes to provide relief to patients with severe pain. Medications can also come from the slimy, the legless, and the undersea creatures of the world. This is not the hogwash of fake pills peddled to unsuspecting victims. This is science. Healing in its truest form.

When Scottish physician Alexander Fleming reflected later in his life on penicillin, derived from a fungus, he said, "When I woke up just after dawn on September 28, 1928, I certainly didn't plan to revolutionize all medicine by discovering the world's first antibiotic, or bacteria killer. But I suppose that was exactly what I did."[85] Penicillin and antibiotics like it have since saved millions of lives. Other examples include the drug Taxol, which is derived from the lush Pacific yew evergreen and is used to treat lung, ovarian, and breast cancer. Captopril, a common antihypertensive medication, is from the venom of a pit viper. Extracts of the aromatic sweet wormwood plant are used to treat malaria. Cephalosporin, also from a fungus, is used for bacterial infections. Sea sponges have inspired medications for HIV (AZT), late-stage breast cancer (Halaven), and leukemia (Ara-C).

Ziconotide, a medication for severe chronic pain, comes from the sea-dwelling cone snail *Conus magus*, or "magical cone." These snails may not look like much with their splotchy shell and deliciously described "chocolate interrupted lines,"[86] but they shoot out a dart-like, venomous harpoon and send their prey into drug-induced chaos. The harpoon is a modified tooth that is formed inside the snail's mouth—a homegrown weapon at its disposal. Once it has paralyzed its fish, the snail retracts its harpoon, swallows its doomed prey whole, and regurgitates any indigestible parts like spines and scales, along with its harpoon tooth. At any one time, a cone snail may have as many as twenty of these harpoons, loaded and ready to fire off another round if prey swims by. If the snail needs to restock its weaponry, it grows some more. Its venom is a potent painkiller and not to be messed with—one thousand times as powerful as morphine.

Impressive as this list is, it doesn't even begin to grasp the potential power of plants that contain psychedelics like psilocybin, LSD, MDMA, ayahuasca, and ibogaine. As much as we may sometimes fear nature and its catacombs, it is also our cradle. And yet, just as we are coming to understand and appreciate the importance of this diversity among Earth's creatures, it is slipping through our fingers. Modern extinction rates are estimated to be one hundred to one thousand times greater than the natural baseline rate. Some 26 percent of known mammals, 40 percent of amphibians, 33 percent of reef-forming corals, and 14 percent of birds are threatened with extinction, according to the International Union for Conservation of Nature. Conservative estimates suggest we are losing one important source of drug every two years.

Just as we have gained the technological know-how to sequence this molecular diversity, no longer harming populations to collect its medicinal qualities, we risk wiping them out. For some species, it is already too late. We are, as Professor John Malone of the University of Connecticut says, "destroying the best library in the world."[87]

First Impressions

IF WE WANDER Earth's metaphorical library to the section on large venomous reptiles and flip open the pages of a book on Gila monsters (pronounced HEE-luh), it's easy to see why these creatures were destined to be viewed with unkind eyes. The Gila's scientific title is *Heloderma suspectum* (from the Greek *helos* for "nail stud" and *derma* for "skin"), which conjures something secretive and perverse. This is no doubt the result of the curse of first impressions: these two-foot lizards roam the hardy creosote thickets of the southwestern United States and are covered in a sinister patchwork of black and orange beaded scales. They are, by all accounts, formidable in size compared with their skittish lizard peers.

Described by paleontologists as "living fossils," Gila monsters hark back to the Age of Dinosaurs, when winged reptiles took to the skies and dagger-toothed giants prowled the land. Fossils of their kind have been found alongside tyrannosaurids in what is now Utah and even in Mongolia, where they were unearthed near nests of the neck-frilled protoceratops, whose young they possibly fed upon. Their spooky looks have certainly hoodwinked humans. For more than a century, Gilas have electrified headlines: It is "the most deadly reptile in all the world, not excepting the cobra of India or the Staked Plains rattlesnake of Texas," wrote the *Newberry Herald* in 1890. The *San Francisco Chronicle* merely insulted it, calling it "Ugly as Sin Itself" (1893).[88] The newspapers reflected the mood of the time—until a man named Dr. George Goodfellow put their slurs about the Gila's death-dealing bite to the test.

Goodfellow was an American doctor in one of the Old West's roughest, toughest, and most bloody towns: Tombstone, Arizona. When he stepped through the saloon's batwing doors into the desert sun in 1888, everyone recognized the mustachioed gentleman. His dark hair was parted down the middle with scalpel precision and a gun dangled from his hip. He had earned himself a reputation as a "gunfighter's surgeon," and there was ample opportunity to practice his trade in the lawless town, where disputes were regularly settled with a slug from a revolver.[89] The dusty frontier was home to outlaws, lawmen, cowboys, miners, merchants, brothels, get-rich-quick schemes, and whiskey that was used as both a sterilizer and a drink. It was also home to a feared monster lizard.

Unfortunately, Goodfellow wasn't doing so well. If you were in the saloon that day and had an eye for detail, you'd have caught sight of his pale face and bandaged finger. Under the dressing, the skin was swollen crimson and pitted with dark bite marks. For the last five days, the doctor had been bed-bound thanks to his own curiosity. Goodfellow had many interests, among them philosophy, geology, and creatures with undue reputations. In regard to the latter, and in reaction to newspapers that claimed people were dying from the Gila's venom, he'd caught one of the creatures the week before—this one the size of a cowboy boot—and provoked it to bite his finger.

The reptile's chomp, he discovered, is more of an excruciating chew than a quick crunch, but he had proved his point—he'd survived the Gila bite, only a tad worse for wear. He proceeded to write a counterargument in *Scientific American*, calling the Gila's deadly reputation "the remnant of primitive man's antagonism to all creepy things."[90] His observations, confirmed years later by research, show another side to the monsters; they are shy and prefer life in the slow lane, underground and away from our prying eyes. Here they can stay for months, their bodies feeding off fat stashed in their sausage-shaped tails. It's hardly the average portrait of a killer.

Gilas may not be murderers, but they do have the tenacity of a desert survivor. Dale DeNardo, a biology professor at Arizona State University, likens their bite to being hit by a hammer every ten seconds for forty-five minutes. Similar to Goodfellow, though, DeNardo has a soft spot for the reclusive reptiles. Tucked behind a security door in his laboratory are all sorts of lizards and snakes. It's the Gila monsters, however, that he's been studying for the last twenty years. It's tempting to say DeNardo has learned all there is to know about the creatures, but he's still at it, putting them in watertight diapers to see if they sweat from their cloaca—the opening where urine and feces are eliminated (spoiler: they do)—and raising Gila monsters that are now older than most of his students. "One of my interests in them early on was reading work written by scientists about how poorly adapted, or even maladapted, they are for living in the Sonoran Desert," says DeNardo. "To me, that didn't make much sense. They've been in the Sonoran Desert for as long as there has been the Sonoran Desert. We're missing something here."[91]

At first, Gila monsters appear to be a random mishmash of animal parts: they are venomous, like a snake, and keep a chunk of fat in one place, like camels. Gilas can go without food and water for up to four months by stockpiling fat in their tails, which can swell to the size of a hog sausage or waste away to a slender finger, a testament to its fortunes and the season. The lizard can also store diluted urine to stave off dehydration, like a desert tortoise, its bladder filling to nearly 22 percent of the animal's weight, the equivalent of a hiker schlepping around a thirty-pound water bottle. Essentially, "they threw the lizard rule book away," says DeNardo.

In one experiment in the late eighties, a Gila munched on four bald newborn cottontail rabbits. This may not seem like an incredible feat of biology—at least not until you realize the lizard ate one-third of its body mass in a single meal. If we compare that with champion eater Takeru Kobayashi from Japan—as Professor Daniel D. Beck from Central Washington University did—the Gila monster's feat is, quantitatively speaking, more impressive. In the summer of 2002, Kobayashi polished off a record-breaking 50.5 hot dogs in twelve minutes, consuming more than 8,000 calories, quadruple the amount an average person needs in a day. By Beck's calculations, Kobayashi could sustain his usual energy for four days. The Gila could last for nearly four months on the four cottontails. When Beck took the metabolic rates of both the reptile and the champion eater into account, he calculated that a Kobayashi-sized Gila monster would be able to sustain itself on one meal of fifty hot dogs for over a year. This is simply impossible for humans. We must constantly chase the ephemeral satisfaction of food, knowing that our hunger will soon return. As Ralph Waldo Emerson wrote, "I can reason down or deny everything, except this perpetual Belly; feed he must and will, and I cannot make him respectable."[92]

How, then, does the Gila feed its belly so intermittently? If we squeeze our way inside the lizard's mouth, we'll find that its saliva is a double-punch cocktail. The "venomous portion of the spit is mainly there to defend the Gila monster," says DeNardo. "This is supported by the fact that a Gila monster bite is instantaneously painful." The agony distracts the attacker, giving the lizard time to scamper away into the bushes.

What we didn't discover in the Gilas' saliva until much later than the venom is the second part of the cocktail. Inside lies the secret to how they keep their blood sugar remarkably stable as they swing between starvation and binging—a master stroke that eludes even the hardiest of patients with diabetes.

Gut Instincts

TOMBSTONE COULDN'T BE more different from the Bronx, New York, except that the Bronx, too, brings people together from all walks of life. When Dr. John Eng, an endocrinologist at the Bronx Veterans Administration Medical Center, heard of the Gila monster's venom engorging the pancreas of its victims, where insulin is synthesized, he made it his mission to study this creature that others feared.

In the 1980s, Dr. Eng was working in the lab of Nobel Prize winner Dr. Rosalyn S. Yalow, who won the prestigious award for devising a technique called radioimmunoassay, which measures the amount of hormones, enzymes, and other substances in a fluid. Her technique is so sensitive, it's said to be capable of identifying a teaspoon of sugar in a lake 62 miles (100 kilometers) long, 62 miles (100 kilometers) wide, and 30 feet (48 kilometers) deep. The first hormone Yalow's team explored is one believed to have originated more than a billion years ago: insulin. Yalow was particularly invested in this line of research because her husband was a diabetic. Diabetes mellitus is a disease that pummels the pancreas, an organ snugged between the intestine and the spine, where insulin is produced.

Before insulin was discovered in 1921, patients with diabetes didn't live long. The best that doctors could prescribe was a diet: eat low-carbohydrate meals—and not much at that—and you may (or may not) live for a couple more years. It was a dreary diagnosis. Mentions of the condition date as far back as Egyptian medical texts written in 1552 BCE (known as the Ebers Papyrus) and to ancient Indian and Chinese texts. The word *diabetes* comes from an ancient Greek word meaning "a passer through," and *mellitus* comes from a Latin word meaning "sweetened with honey." Together, *diabetes mellitus* refers to the sweet taste of urine produced by those with the condition.

By the early 1900s, doctors knew that something in the pancreas gland was responsible for diabetes. This knowledge was based on studies by German researchers, who found that dogs without the gland experienced diabetic symptoms and perished soon afterward. In 1922, a thirty-year-old orthopedic surgeon named Frederick Banting and his assistant Charles Herbert Best, a twenty-one-year-old medical student, revolutionized our understanding of the disease. Banting was raised on a small Ontario farm and almost became a minister before interrupting his divinity studies for medicine. The switch was fortuitous, as later a fourteen-year-old boy named Leonard Thompson was dying from type 1 diabetes, and Banting was his only hope. Thompson's weight dropped to a mere 65 pounds, his energy levels plummeted, and his blood-sugar levels soared wildly.

In previous experiments, patients had been fed freshly minced sheep pancreas, but to no effect. Banting wondered whether a better solution would be to tie off a cow's pancreatic duct to isolate the gland's internal secretions, a yet-to-be-identified substance, and inject the secretion into patients (cattle pancreases were more widely available than those of dogs and sheep, since many were already slaughtered for food). He proposed this idea to Professor J. J. R. Macleod from the University of Toronto, who was unimpressed but begrudgingly allowed Banting to use his laboratory. Banting went to work, tying off the pancreatic ducts, extracting the mysterious substance, and injecting the secretions into a diabetic dog named Marjorie. After an hour, her illness improved. Banting named the secretions *isletin* because they were extracted from islet cells in the pancreas. Professor Macleod, who had initially rebuked Banting, now devoted his lab to further investigation of the secretion, changing the name from *isletin* to *insulin*, Latin for "island" (first coined by physician Edward Albert Sharpey-Schafer, who theorized there was a mysterious substance produced in the pancreas that might cause diabetes).

Not long after, in a desperate bid to save his child, Thompson's father agreed to let the doctors inject his son with a purified version of the insulin. On January 11, 1922, Leonard Thompson became the first person in the world to receive an insulin injection. Within twenty-four hours, Thompson began to feel more alert, but after a couple of days his symptoms returned, so he was injected again. Thompson again felt better, and his levels stabilized. The "medical miracle" made front-page news around the world. In 1923, Banting and his colleague J. J. R. Macleod won the Nobel Prize in Medicine for their discovery of insulin (sharing the prize money with two other colleagues). Although a life-saving treatment, it was far from a perfect medication; insulin from cattle and pigs triggered an allergic reaction in many patients. Only in 1978 did scientists genetically engineer "human" insulin using *E. coli* bacteria, shrinking the likelihood of an allergy to the medication. By taking insulin injections, Thompson lived for thirteen more years, before dying of pneumonia (likely as a complication of his diabetes) at the age of twenty-seven.

Nowadays, more than 420 million people worldwide suffer from diabetes, over three-fourths of whom have the type 2 variety, which is caused by a combination of lifestyle factors and genetics, with unhealthy food choices and a lack of exercise playing a prodigious role. The saying "You are what you eat" isn't *exactly* true, but there's definitely something to it. All of us are alive because of millions of chemical reactions taking place inside our bodies every second. The world is ruled by chemicals, including all that we drink, eat, and breathe. These chemicals are assembled into molecules that structure themselves in space, and then move through that space to interact with billions of other molecules doing the same thing. Some become a part of us to energize our cells, power our brain, repair cells, or create new cells to replace previous generations. We are, it is safe to say, a collection of innumerable parts, assimilating chemicals into our bodies as we move through the world.

When we absorb the chemistry of our food, our bodies start to break down that food's nutrients in our stomach and turn them into fuel. This is the first step in powering our metabolism, but it's only half the battle. Most of the cells in our body need glucose (or sugar, for simplicity's sake) and fats for energy. You can think of sugar as energy parcels that sail through the bloodstream, a vast system of vessels that's over 60,000 miles long—enough to wrap around the world more than twice. The sugar somehow needs to make the move from our bloodstream into our cells. If there's nothing poised to grab the sugar parcels as they sail past, the nutrients will be lost. That's where insulin comes in. The hormone is vital to a healthy body. After the sugar in our food is broken down, insulin is released from the pancreas and embarks on a journey through the bloodstream, where it attaches to cells and helps them snatch the sugar as it floats by. To use an analogy, the cells are like docked ships and the insulin is the fishermen onboard, trying to snag fuel for the crew. Once in our cells, the sugar can be converted into energy for use now or stored for later.

That's how the body works for a person without diabetes. For those with the disease, either not enough insulin is made or the body doesn't use it properly. Type 1 diabetic patients don't make enough insulin, so we give them insulin injections to replace what they lack. For type 2 diabetic patients, it's more complicated, because the condition is more of a lipid, or fat, issue. These patients can make insulin, but when extra fat settles in their pancreas and muscle cells, the fat disrupts the normal operations of these body parts. Fat clogs up the pancreas's ability to secrete insulin in a timely manner; equally, the insulin doesn't bind well to their muscle receptors. If insulin struggles to grab the sugar parcels, and too much sugar is left in the bloodstream, the result is high blood-sugar levels. And too much sugar in the blood for too long can damage the vessels, increasing a person's risk for heart disease, stroke, kidney disease, vision problems, and more. The situation is exacerbated by the insidiously addictive nature of sugar. Grab random packages and bottles from grocery aisles and you'll find added refined sugar in almost everything, from pasta sauces and breads to salad dressings and granola bars.

The average American consumes between seventeen and twenty-two teaspoons of sugar a day, or fifty-seven pounds a year. The recommended amount is no more than six teaspoons of added sugar per day for women and nine for men. We act like hummingbirds, chasing sweetness at every turn. Knowing this, it's tempting to boil type 2 diabetes down to poor diet choices, but sugar is an invisible, hard-to-spot enemy. Our desire to gorge on sweets and fatty foods is a good survival strategy—at least it was for our hunter-gatherer forebears. In the hot African savanna, high-calorie sweets were few and far between, and when we found them they were packaged as ripe fruits or nuts. It made sense to eat as many as possible before baboons or other creatures discovered the sugary or fatty trove too. Sugar and fat are efficient sources of energy, and we're genetically hardwired to enjoy them; they helped us to survive tough times hundreds of thousands of years ago. Today, though, our genes are stacked against us in the face of fast food, cheap calories, and twenty-four-hour buffets.

This means type 2 diabetes has risen to the seventh-leading cause of death in the United States; it's a disease the World Health Organization calls a "slow-motion catastrophe."[93] Up to 50 percent of people with diabetes will experience nerve damage to their limbs, leading to sores, ulcers, and even amputations. Visual impairments will develop in 80 percent of patients within fifteen years, and of those about 2 percent will become blind. Our modern world may allow us to have food delivered to our doorsteps, but our DNA is still stuck in our hunter-gatherer past.

"Medications to treat type 2 diabetes are saviors at the eleventh hour of disease progression," says Judith Kalinyak, a doctor in endocrinology and internal medicine who worked as an attending in the diabetes clinic at Stanford University. "A better choice is prevention, not medication."[94] The human body is an elegant but complex system, and any adjustment to its internal chemical balance has the potential to tilt another system off-balance. Using medication to treat chronic diabetes isn't like flipping an off switch back into the "on" position. It is an attempt to find an internal balance that allows for a better quality of life with as few adverse effects from the drug as possible. Unlike engineers, who can build a new device from scratch, doctors must work within the limits of the human body's design and make tweaks to existing parts.

Knowing all this, Dr. John Eng set out to find a treatment for patients with type 2 diabetes by looking inside the jaws of a monster.

Monster Medicine

IN THE LATE 1980S, Dr. Eng was sifting through the pages of recent studies by the National Institutes of Health when he discovered that the venom of certain snakes and lizards increases the size of a victim's pancreas. This suggested that something in the venom was overstimulating the pancreas—potentially good news when it comes to insulin production. Dr. Eng first focused his attention on the Mexican beaded lizard and purchased its venom to study. In lab tests, however, the hormone in this lizard's spit sent an animal's blood pressure plummeting to a dangerous degree. Eng promptly dropped that line of research and turned instead to the Gila monster, a creature that can slow down its metabolism after long periods without food. Here he hit gold.

After a big meal, a hormone in the Gila's blood spikes thirtyfold to control the surge of sugar in the body, helping the lizard to safely digest food over many months. The peptide hormone, which Dr. Eng called Exendin-4, triggers the release of insulin from the pancreas—just what patients with type 2 diabetes need. But that wasn't the end of Dr. Eng's research. Not even close. He needed to compare the Gila's hormones with those in our bodies. The more similar the hormones are, the higher the chance the Gila's version will work in humans too. In a beautiful twist, it turns out that Exendin-4 is intriguingly similar to a human intestinal hormone called glucagon-like peptide 1 (GLP-1).

Think of GLP-1, which our bodies produce within minutes of eating, as messengers sprinting to give the pancreas the heads-up that it needs to start releasing insulin. Typically, people have two small hormones that control their insulin release: GIP and GLP-1. In patients with type 2 diabetes, the GIP is broken or not functioning, while GLP-1 function remains. The idea scientists had was to stimulate the healthy hormones to compensate for the non-functioning ones. Scientists have tried to inject extra GLP-1 into patients to solicit more insulin from the pancreas, but it lasts only minutes before the body breaks it down again—hence the need for something that spurs faster insulin release, something similar to GLP-1 but long-acting. The Gila's hormone is just that. In type 2 diabetic patients, insulin release is slow and food absorption too fast, meaning the insulin doesn't have time to snag the sugar and help get it into cells. In human clinical trials of patients with type 2 diabetes, researchers found that Exendin-4 slows the time it takes for food to empty from the stomach and triggers the production of insulin from the pancreas when sugar levels are high in the blood. Exendin-4 brings the body back into balance, allowing insulin release to match sugar absorption.

And yet, for all its promise, Exendin-4 was slow to see the light of day. The U.S. Department of Veterans Affairs, where Dr. Eng worked at the time of the discovery, couldn't patent Exendin-4 because it wasn't related to veterans. This was a blow: no pharmaceutical company would take a bet on a drug if it wasn't patent protected. It takes lots of time and money for a drug to go through three phases of clinical trials in order to ultimately reach the FDA for approval. If an idea isn't patent protected, it's too easy for all of that effort to be poached by other companies hunting for a similar prescription. Dr. Eng believed so much in Exendin-4—"his fifth child"—that he filed for a patent on his own in 1993.[95] Two years later, he received patent 5,424,286 from the United States Patent and Trademark Office. He then met with drug companies large and small, including Eli Lilly and Company, one of the big pharmaceutical guns in town, determined to hook them on a reptile hormone. Still he got no bites...

A Young Start

MORE THAN A year after he received his patent, Dr. Eng stood next to his poster in a San Francisco conference hall with a lanyard looped around his neck, as did many other presenters at the annual meeting of the American Diabetes Association that day. Hundreds of people walked by Dr. Eng before a man named Andrew Young stopped and peered at the poster. At the time, Young was head of physiology at Amylin Pharmaceuticals, a biotechnology start-up researching peptide hormones in the hopes of treating type 2 diabetes. But Amylin was at its wits' end over the speed with which GLP-1 degrades. They just couldn't get it to last—and here was a man standing politely in front of a poster with a solution to that very problem. Immediately intrigued, Young rang the company to share Dr. Eng's discovery.

Within a month, Dr. Eng found himself seated at the Amylin table, along with everyone who counted. During the few weeks that had passed between the conference and that meeting, Amylin had verified Exendin-4's traits and, at the same time, discovered properties even Eng did not know about. He licensed his discovery to Amylin for a sum the *New York Times* reported as far less than $1 million ("I still have to keep my daytime job," he told them[96]). By 1998, two years into research and development, Amylin's stock had crashed and the company had shrunk from three hundred employees to just thirty-seven. Yet the company was betting on the future success of Exendin-4 as its saving grace.

In 2002, salvation came in the form an ironic twist. Eli Lilly, which had initially turned down the chance to develop the drug, pumped $325 million into Amylin to keep the drug alive and jointly market it. Finally, in 2005—twelve years after the patent process began—the synthetic lizard-inspired medication exenatide (brand name Byetta) was approved for use in the United States. (It's important to note that the medication is inspired by the mechanisms of action found in the lizard's spit; there are no farms "milking" Gila monsters of their venom.) In 2006, the European Commission also granted approval. And things took off from there. The type 2 diabetes drug proved so popular that, a year in, there was a shortage of the pen cartridges used to administer the medication. To do this, people with diabetes inject the clear liquid under the skin of their thigh, abdomen, or upper arm for five seconds in twice-daily doses to help control blood-sugar levels. A prefilled pen contains sixty doses of Byetta, for thirty days of use. It is often prescribed with other medications to help lower blood-sugar levels.

Much of the drug's popularity is due to one of its side effects: weight loss. In a clinical trial of more than five hundred participants over a period of six months, Byetta users lost on average five pounds, while insulin patients gained four pounds. A smaller trial found a twelve-pound weight loss after two years on Byetta (or "Lizzie," as patients fondly call the lizard-inspired medication). It isn't enough for patients to become lean machines, but it is an improvement on the common side effect of weight gain for other diabetes medications. The weight results vary, and the most common side effect for patients is nausea, which for some is intolerable enough to stop the medication altogether. Still, according to a 2018 report, the Byetta family of medications made some $700 million in revenue that year.

Scorning Our Saviors

RESEARCHERS HAVEN'T STOPPED at the Gila monster in the search for potential diabetes medications. A pale cave-dwelling fish called the Mexican tetra, whose eyes waste away as they age into adulthood, may inspire the next generation of meds, according to geneticists at Harvard Medical School. In what sounds like the stuff of a Brothers Grimm fairy tale, the fish repurposes its empty eye sockets as storage pockets for clumps of fat. But that's not even the crazy part, at least not for scientists. If you dissect the tiny swimmers, you find that their organs are surrounded by fat. If this were to happen in humans, it would inflame our body's organs and narrow our blood vessels, resulting in high blood pressure and increased risk of diseases like diabetes, heart disease, stroke, and high cholesterol. In a weird twist of evolution, the fish's blood sugar spikes and crashes, similar to patients with type 2 diabetes, and yet they experience no health issues. In fact, the cave fish has the same genetic mutation as people with Rabson-Mendenhall syndrome, a rare diabetes-like disorder in babies that leads to death within a few years. How do the fish survive? The secret remains a mystery, but it could harbor the clue we need to treat similar conditions in humans.

Other nature-inspired stories lurking in our medicine cabinets include some nine marine-derived drugs that have been clinically approved as of 2020 to treat cancer patients, according to Midwestern University. An auspicious compound in a glass sponge (*Aphrocallistes beatrix*) may also be harnessed to fight pancreatic and breast cancer. Maggots, or fly larvae, can twist even the most hardened face into disgust, but the larvae's love of feasting on rotting flesh has been used to help those who suffer from chronic wounds and infections. Around 6.7 million people in the U.S. have chronic wounds that, for various reasons, will not heal. Not all of these patients will need maggot therapy, but in 2008 alone, maggots were used some 50,000 times worldwide to treat patients. Incredibly, the smell of death is so attractive to mother flies that they can sniff out decomposition up to ten miles away and can even dig through six feet of dirt to get at a dead body. In hospitals, disinfected maggots are placed on a wound and bandaged for up to seventy-two hours, to keep the critters both munching on the injury and out of sight from potentially revolted patients. As the larvae squirm in the wretched tissue, rough skin and hooks in their mouth scrape away dead flesh. Digestive enzymes and antimicrobials ooze out to help the maggots consume the dying flesh, leaving the patient's healthy tissue virtually unscathed.

Then there are synthetic calcitonin drugs spawned by salmon. Although humans make calcitonin in their thyroid gland to regulate their calcium levels, the salmon's version is fifty times more potent than our own. The FDA has approved laboratory drugs that mimic the salmon's calcitonin for the treatment of osteoporosis, or bone loss, in women who are beyond five years postmenopausal and for patients with Paget's disease, a disorder that disrupts bone renewal and causes bones to be weaker than normal.

A protein in the venom of the southeastern pygmy rattlesnake that leaves prey bleeding out and unable to clot was made into an antiplatelet drug (called eptifibatide) for those with advanced heart disease, preventing blood clots from forming and causing a heart attack. Similarly, tirofiban, a blood thinner that staves off blood clots, comes from the saw-scaled viper, a species listed among the "Big Four" dangerous snakes in South Asia.

Leeches may seem like an archaic form of healing, but their exceptional talents have crept back into modern surgical rooms. Leech saliva contains hirudin, an anticoagulant that keeps blood flowing to damaged areas, encouraging new veins to grow, which can be helpful, for example, in skin grafts or reconstructive surgery. Apart from the cringe factor, leeches are a fairly painless treatment option. A nurse pulls the slippery parasites out of a container using forceps and directs their heads toward the appropriate wound site, where they latch on to a patient's tissue and suck their blood. A leech's mouth is lined with three jaws and dozens to hundreds of teeth. In the hospital setting, a leech is left to feed for thirty to ninety minutes and releases its grip when it is bloated with blood. The bloodsucking creatures are not slugs, reptiles, or insects. The leech belongs to the Annelida phylum, a category of segmented worms that includes 22,000 species. Leeches have nine pairs of testes and are often said to have thirty-two brains, though it's more precise to say their noggin is spread out over thirty-two parts of their body. In 2004, the FDA officially approved leeches as a respectable therapy to use in medicine.

There's also lanolin, a yellow waxy substance secreted from the sebaceous glands of wool-bearing animals. Crude lanolin makes up some 5 to 25 percent of the weight of freshly shorn wool, and it is used in everything from cosmetics to moisturizers to protective baby skin care, as well as to treat nipples that are sore and cracked from breastfeeding.

These medicines are worlds away from the fake drugs hawked on the black market, which has contributed to record extinctions and endangered species. Thanks to modern technologies, we can tread more softly on nature in our pursuit of knowledge. These days, scientists sequence and clone a species' DNA to study its molecular makeup, rather than harvesting large numbers for investigation.

Animal toxins, too, are a rich source of active molecules. Most venoms and poisons have evolved over hundreds of millions of years and possess superb stability, potency, speed, and specificity, hitting specific molecular targets. There are currently eleven approved toxin-based molecules on the market, one from cone snails, two from lizards, two from leeches, and six from snakes. And then there is the geographer cone snail that has even weaponized insulin. The tropical sea snail unleashes a cloud of insulin into the water to plunge a fish's blood sugar and send it into hypoglycemic shock. Scientists call the chemical cloud a "nirvana cabal" whereby the chemical invades the fish's gills and floods their bloodstream. New insights into the structure of this insulin could one day lead to a faster-acting treatment for diabetes.

Although we often liken our bodies to machinery, a better analogy might be a complex chemical system—a distinction all living things on this planet share. The question of how these chemical parts come together to form life remains a fundamental mystery. While the Gila monster's tenacity is evident in its evolutionary history, its numbers are falling due to human predation and habitat destruction. It is listed as "near threatened" on the International Union for Conservation of Nature's Red List.

Globally, many medicinal plants are on the verge of extinction or already are wiped out. Some 17 percent of the Amazon has been lost in the last fifty years. The same is true for our underwater hubs of diversity: coral reefs. Deforestation and unregulated extraction of forest resources are major threats to these treasured hot spots. Preservation can, and has proven to, make a difference—if we care enough to tread lightly on a fragile planet.

In 1952, the Gila became the first venomous animal in North America to be granted legal protection status. For many patients with diabetes, this was a fortuitous decision indeed. Their lives have been forever altered by a man curious about a shy reptile hiding in the creosote thickets of the southwestern United States.

11

BUMPS ARE BEAUTIFUL

Whale Warts Inspire Energy-Saving Fans

"The rules of life are very different in this pitch-dark, chilled, high-pressure environment, leading evolution down different paths."
HELEN SCALES, author and marine biologist

IN 1815, *ESSEX* was a ripe eighteen years of age. It was a masculine name for a round-bottomed lady, but captains loved her anyway. She was lucky, it was said. Unfortunately, *Essex* didn't smell as lovely as she looked. Locals liked to joke that the ship stank so bad you could get a waft of her before you saw her. Her stench was a foul consequence of her trade. She was a hunter of the ocean's leviathan, the grand and ever-mesmeric whale.

When seamen weren't steering her into port, she was drifting in the Pacific Ocean. On a typical summer day, seamen cleaned her timbers and scampered up her masts to scan the horizon for a telltale spout. The heat beat down on them, their cheeks ruddy from a life spent under an intense sun. Before them, an endless horizon of glassy blue and a metronomic sea swell, flecks of light dancing on the waves like scintillant fairies. Managing monotony was a time-honored job aboard these ships, which could go for weeks without seeing a whale. In the nineteenth century, whales were commodities, and it was a one-sided affair: conqueror and conquered.

"Thar she blows!" a man shouted from the mast.

A spout of salty, snotty air blasted high into the air, a rainbow gleaming in the drops. The men rushed to lower a whaleboat from the ship's crane-like davits. The whaleboat was twenty-five feet long and the prey they stalked double that size. Their quarry were docile creatures, but a flick of the tail could still splinter wood, flinging boat shrapnel into the air and tossing the men into the cold sea, sometimes miles from the mother ship. A whaleboat splashed into the water and six men clambered inside. They rowed in unison, battling the waves to maneuver themselves within striking distance of the whale. As they drifted mere feet from the mighty creature, the harpooner seized his whale iron and speared the whale.

Whale hunting was a brutal process. A harpoon strike wasn't meant to kill the whale but to capture it, the blunt end of the harpoon attached to the boat by a coil of line. The wounded whale could drag them for miles on what the men called the "Nantucket sleighride," the boat slapping against the waves as they sped toward the horizon at a heart-pounding speed of up to 20 miles (32 kilometers) per hour. After the whale exhausted all its energy, too weary to continue any longer, a shipmate would lurch forward and lance the giant with a sharp knife mounted on a pole. He would continue until blood spouted from the whale's blowhole and the men shouted "Chimney's afire!" The following days were spent slicing the blubber as thinly as possible and tossing it into burning copper cauldrons. These slices were colloquially known as "bible leaves," and they yielded the crew gallons of oil.

Early whaling efforts focused on right whales, humpbacks, and sperm whales, whose waxy spermaceti flickered candlelight into our homes, burning longer and brighter than any beeswax candle. Whale oil was also a prized ingredient in industrial soaps and lubricants, while baleen was used to cinch and shape corsets, stiffening a woman's movement. A whale's ambergris, a waxy lump formed in their intestines due to the itch of a squid beak, made perfumes linger longer on the skin. The Basques were arguably one of the first peoples to make whaling a business. They erected stone towers along the coast of Spain and France to watch for right whales. From the sidelines, the Catholic Church played a notable role. According to historian Mark Kurlansky, the medieval church imposed restrictions on holy days, prohibiting sexual intercourse and the eating of flesh. Red-blooded meat was "hot" and thus associated with sin. The flesh of creatures that lived in the water, however, was "cold" and couldn't excite passion in the body, so whales and cod were exempted from holy-day restrictions. (We now know this is biologically false, as whales are not fish but warm-blooded mammals that swim in cold seas.)

Whales were global forces of nature. In Korea, the Bangudae Petroglyphs from 6000 to 1000 BCE feature scenes of whales breaching out of the water and mother whales with babies on their backs. In Russia, above the Arctic Circle in a place so remote the region is only accessible by helicopter, tribes from 2,000 years ago carved images into the rocks of people hunting whales alongside dancing women with large hallucinogenic mushrooms on their heads. In Balls Head, Australia, there is a twenty-foot-long Aboriginal engraving of a man lying on top of or inside a whale, possibly to cure himself of an illness or to evoke Aboriginal myths about a human swallowed by a whale. Aristotle wondered whether whales were fish or mammals since they bear live young like mammals (we now know some fish do give birth to live young, their eggs hatching inside their bodies before delivery).

Even Charles Darwin took his turn, speculating in *On the Origin of Species* that a land animal that captures fish like a black bear could, by natural selection, grow "more and more aquatic in their structure and habits, with larger and larger mouths, till a creature was produced as monstrous as a whale."[97] The ridicule he received prompted him to edit this passage in subsequent editions of his book. Darwin had the right idea, though—just the wrong species. He should have been looking at hippopotamuses. Both whales and hippos evolved from four-legged, hoofed ancestors that lived on land some 50 million years ago. Whales are one of the rarities whose ancestors migrated out of the sea and then back into it. The smoking gun was a fossil dug out of an ancient seabed in Pakistan named *Ambulocetus natans*, which means "swimming walking-whale." The creature had four limbs for walking on land, a bit like a sea lion, and hunted in the water by kicking its flappy, webbed feet.

For hundreds of years we've studied whales, hunted them, dissected them, and trapped them in aquariums for all to see. We've studied their fatty pink milk, thick like toothpaste from a diet of 3,000 pounds of pink-hued krill every day, and we've pushed them to the brink of extinction. It's tempting to assume we know all there is to know. And yet, these creatures can still surprise even the most expert among us.

Dr. Frank E. Fish (yes, that's really his name) is a professor of biology at West Chester University in Pennsylvania, with a PhD in zoology. He leads the campus's Liquid Life Lab, exploring how sea creatures with bones swim—from dolphins, otters, and platypuses to hippopotamuses and frogfish. He wants to know everything there is to know about how they move underwater. How do alligators swim so fast with such stubby arms? How do the color-changing frogfish move by gulping water? Fish has taught for the last forty years, strolling past dozens of doors on campus to one door in particular, tucked away in the basement of a building. The door opens to inflatable toy dolphins and whales, and nearby, a freezer stores the frozen flesh of a long-dead creature. During our interview, Fish wears a dark T-shirt with a majestic whale print front and center. It's safe to say that Dr. Fish knows a lot about the life in our oceans. And yet, one day in the early 1980s, he found himself standing in the middle of a gift shop in Quincy Market, Boston, feeling utterly surprised. In front of him was a statue of a humpback whale on a pedestal, twisting in a dive. But something seemed off.

"A former mentor of mine used to comment on artists' paintings and say how realistic they were and [point out] if they forgot anything," says Fish.[98] As an homage to this mentor, Fish inspected the humpback whale with a critical eye—and quickly noticed something wasn't right.

"I'm looking at this figurine and saying, 'This is wrong,'" he says. "There are bumps on the front of the long flipper. I know hydrodynamics. I know aerodynamics. The artist got it wrong."

The bumps, thought Dr. Fish, made no sense, especially on the front of the flipper. The commonly understood laws of hydrodynamics (how objects move through water) and aerodynamics (how objects move through the air) suggest that being streamlined matters most for speed. This is why elite swimmers shave their body hair before meets, and why aircraft wings are made of sleek titanium. Friction creates resistance, and resistance impedes movement and slows speed. Why, then, would a whale with one of the longest migrations on the planet and weighing 70,000 pounds—the equivalent of fourteen SUVs—grow knobs the size of fists on its fins, where being streamlined matters most?

Fish started laughing and the manager came over, wanting to know what the fuss was all about. When Fish told her the artist had gotten it wrong, and that the whale's flipper should be smooth, the woman shook her head. "Oh no, no," she said. "This artist is very, very careful."

The woman handed him a brochure featuring a photograph of a humpback whale. Sure enough, there were nine or so bumps on the leading edge of the flipper. Curiosity piqued, Fish decided to get to the bottom of this mystery. But in order to learn more, he had to find a dead whale to examine. And he wasn't about to kill one. He needed a whale that had died of natural causes. So he called up a friend at the Smithsonian, a place that collects all sorts of critters, and asked if, by any chance, they had a spare humpback whale lying around. After a brief pause, the man on the phone replied, "No, but I'll put you down on the list."

Dr. Fish waited years, experimenting with other creatures, until one day he received a call from his friend at the Smithsonian: "A humpback whale has died and washed up on a beach. They'll allow you to get a flipper. But you've got to go today."

"How big is the whale?" Fish asked.

"Twenty feet," the man replied.

A humpback's flipper is one-third the length of its body. Doing some quick math, Fish calculated this particular flipper at about six feet—and sighed with relief; it would fit in his car. He drove all the way to New Jersey and stepped out onto the beach to find a ten-foot flipper, as tall as an elephant.

"Hold it!" he said, aghast. "You said the whale was twenty feet long!"

The man shrugged. "We made an error. The whale is thirty feet."

Dr. Fish stared at the flipper, as white as sea foam and speckled with dark blotches. There was no way it was going to fit in his car. Frustrated, he spent most of the day cutting the flipper into three chunks, each weighing well over a hundred pounds, and wrapping them in black plastic. He shoved the hefty pieces into the trunk of his little Ford Escort and watched as the rear of his car sank toward the ground. He then drove back to the laboratory in a droopy car, worried police would pull him over and find rotting body parts in his trunk.

"I was in absolute terror that the New Jersey state trooper would ask what was in the back."

Eventually surmising that New Jersey cops regularly have to deal with the mafia, and that a whale fin would be far from the worst thing they'd seen, he took his chances and returned to West Chester University. From there, Dr. Fish took the question of why humpbacks have bumpy flippers and swam with it, finding that even after all these years, the hunchbacks of the underwater world can still blow us out of the water.

Saltwater Acrobats

SCHOLARS SUCH AS Dr. Fish call the fleshy knobs "tubercles," from the Latin *tuberculum*, meaning "tuber," like a bumpy potato from a garden. Old-time whalers called the tubercles on a humpback's top jaw "stove bolts." Either way, the tubercles must either serve a purpose or be the remnants of some vestigial anatomy, since they are present on the whale's fin even in the womb. The bulges are especially curious given that of all the species of whales in the world—and there are eighty-six in all—it is the humpbacks that triumph at tight turns. During one such move, called an "inside loop," the humpback rolls 180 degrees, makes a sharp U-turn, and then lunges at prey. The grizzled whale blows hoops of bubbles during this maneuver in the same way that Gandalf from *The Lord of the Rings* puffs rings of smoke, except the whale's underwater exhalations serve a vital purpose: they cluster disoriented schools of fish into a small circle. Once a whale sounds the feeding call, the pod surges in unison from the dark depths.

Despite their size, humpbacks are stout and meager compared with their cousin the blue whale. Instead, their claim to fame is that fin that so fascinated Dr. Fish. The whales' scientific name, *Megaptera novaeangliae*, means "giant-winged New Englander," a fitting tag considering the humpback's fins are the longest appendages in the world, growing to a prodigious sixteen feet (nearly five meters).

Here, then, was the question that had taken up real estate in Dr. Fish's imagination: If humpbacks are known for their sharp turns, do the bumps on their massive fins somehow help them to chase after prey? At first, some scoffed at the idea. The basic tenets of aerodynamics and hydrodynamics are as ingrained in engineers as the basics of DNA in molecular biologists. Even a small amount of friction can slow the flow of air or water over an object. How, then, could these so-called warts make whales better swimmers? That's like saying acne makes chimpanzees better climbers, isn't it? Well, not exactly.

It's the location of the tubercles that makes a difference. In collaboration with Dr. Laurens Howle of Duke University, Fish performed wind-tunnel tests of scale-model flippers with and without those knobbly edges. Fifteen-foot flippers became two-foot-long models, and the golf-ball-sized tubercles shrank to the size of sunflower seeds.

To the shock of nearly everyone, they discovered that the bumps *do* help humpbacks to swim. The warts swirl the water into small whirlpools called vortices; it's as if they are stirring the water as it passes over the flipper. Without the bumps, one large swirl of water could travel down the length of the flipper and slip over the edge, tipping the whale off-balance as it turns. Essentially, a bumpy fin is more acrobatic than a smooth one because the spaces between the bumps create narrow channels of water flow. This helps the whales to pivot quickly as they hunt fish bolting through the water like silver torpedoes.

If the whale's flippers are tilted too high during a turn, the creature could stall and lose the centripetal force necessary for keeping it angled in a curved path. Its prey would dart away. The bumps increase this angle of attack. An easy way to break this down is to imagine sticking your hand out the window of a moving car, palm facing down and parallel to the road. The push on your hand from the air is called drag, essentially resistance to forward motion. If you tilt your palm to the flow of oncoming wind, you now feel both a push and an upward force, or lift. If you keep increasing the angle, you'll eventually get to a point in which the lift falls off and you get lots of drag, and your hand swings backward. That's the stall point. The angle at which this happens is called the angle of attack. For an aircraft, this typically occurs at about fifteen degrees.

Dr. Fish found that a warty fin can decrease drag by up to 32 percent, increase lift by 9 percent, and avoid stalling at high angles—a set of circumstances called the "tubercle effect." But he wondered if there was more we could gain from this knowledge. Could this effect transcend whales and leapfrog our imagination? Shark scales, for instance, have inspired antibacterial surfaces, and research on schools of fish has given rise to wind farm prototypes. Could we apply our understanding of tubercles to the fields of aerodynamics and hydrodynamics? "When you hear the whirl of a fan and a high-pitched noise, that is because of the wingtip vortex," says Dr. Fish, referring to the swirls of air behind the fan's blade that can produce drag. "So if you can reduce that vortex, you can make it way more efficient and quiet."

Fish and Stephen W. Dewar, an inventor and documentary filmmaker, formed a company and have exclusive rights to Tubercle Technology for wind turbines, fans, compressors, and pumps, though they currently license their designs to other companies. Their first product-to-market invention was an industrial high-volume, low-speed fan, now sold worldwide. The bumpy blades are reported to be 25 percent more efficient, consume 20 percent less electricity, and operate more quietly than standard blades. They have also tested micro-fans to cool computer graphic cards, as well as diesel engine fans.

An American bicycle company added lumpy rims to its wheels to improve stability in high winds, and aircraft designers already use something similar called strakes or leading-edge root extensions. These run from the leading edge of the wing root to the fuselage and are triangular in shape. Used since 1959 on fighter aircrafts, strakes generate vortices and help sustain lift at higher angles of attack. In 2013, a group of German graduate students also put bumps on truck mirrors and tested them in a wind tunnel. The mirrors experienced less drag and turbulence than unmodified mirrors, increasing fuel economy.

But questions remain: If tubercles are such wondrous devices, how come only humpbacks have capitalized on this design in the animal kingdom? Well, it's not just humpbacks that use them. Bumps on the leading edge are rare in nature but not unheard of. During the Paleozoic era, fish of the order Inioperygia had large hook-shaped scales on the leading edges of their pectoral fins. Then there is the scalloped hammerhead shark with a bumpy forehead, called a cephalofoil, that improves hydrodynamic performance. Small tubercles are also found on the dorsal fin of porpoises, and on seal whiskers.

"That helps to modify the flow over the whiskers to get rid of any vibrations," says Dr. Fish. "By having the bumps along the leading edge of these whiskers, it reduces that vibration from the water vortices and allows them to better detect the movements of prey in dark and turbid waters." We could use a similar idea for our underwater robotics, helicopter blades, turbines, and surfboard fins.

Turn out, we still have a lot to learn from these creatures—bumps and all.

Dead Fish Swim Upstream

FROM THE DEPTHS of the universe to inside our very own bodies, vortices are everywhere if you know where to look. We see them in the galactic spirals of cosmic gas, the twisting pillars of tornadoes tearing tree from trunk, and the flowing blood in our very own hearts. They exist in Jupiter's Great Red Spot—a hurricane that's larger than Earth and powered by the swirling fury of a vortex—and the aerial whirls that slip from a bird's feathered wing tips as they tilt in the wind. Microscopic animals called rotifers use wheel-like cilia to create a vortex of water to suck bacterial cells into their mouths. Scientists have even discovered that sperm don't swim with a paddling tail, as widely believed, but use a spiraling motion like a corkscrew to move against the thick flow of vaginal fluid. In more subtle ways, we see spirals in the natural world in the form of snail shells, antelope horns, climbing plants, the double helix of DNA, and pine cones. There's no rotational movement here, just efficient organization—as if the vortices were frozen in time. These examples have sparked some scientists to ask if we can somehow capture the energy of vortices. It's not as outlandish as it first sounds. We already know a creature that does this.

Enter the humble trout, the sheen of their pink and orange bodies enchanting many a fisherman, including writer John Gierach, who once asked, "How can so much color and vibrancy be generated by clear water, gray rocks, and brown bugs? Trout are among those creatures who are a hell of a lot prettier than they need to be. They can get you wondering about the hidden workings of reality."[99] This is specially so when it comes to their expertise in exploiting the twisting power of a vortex to conserve energy.

One of the foremost experts in this line of research is biology professor James (Jimmy) Liao and his team from MIT and Harvard. Liao is one of those individuals who seems predestined for a career with fish. When he was two years old, he began fishing in a duck pond at Prospect Park in New York City; soon after, he turned his bathtub into an occasional aquatic zoo, and he preserved fish skins in his basement during a fourth-grade bout of interest in taxidermy. His dad was an avid angler and his parents managed a sushi restaurant; inspiration was served on a veritable plate.

Thankfully, Liao eventually upgraded from his parents' tub to a laboratory tank at Harvard University. Here he found himself in the lab of George Lauder, who was among the first to pioneer an adaptation of a technique called particle image velocimetry. This tactic gave Lauder a way to visualize how water flows behind a fish. What kind of trace do their tails leave in the water? Can we map it as accurately as a footprint in mud? With more than 34,000 species of fish, there is a tremendous diversity of tail and fin shapes; some have forked tails while others have square, mast-like tails. Looking at the water flow behind a fish's rudder can provide clues to its function.

As a fisherman, Liao's questions were tailored to what anglers see in rivers and streams. For example, how does a trout remain still in currents strong enough to knock a human down? With only a flick of a fin, they remain almost motionless even as water gushes by at tremendous speeds. Trout seek rest stops in rivers behind obstacles like rocks to shield themselves from the current; it is the equivalent of standing behind a tree to shield yourself from high winds. However, there's more to it than just protected nooks. If you frighten trout, they dart upstream. Almost always, the creature escapes *against* the current, not downstream with the flow of water, as would seem a faster getaway. Somehow, the trout uses the river to squeeze its body like a bar of soap and torpedo forward.

To understand why, Liao concocted a "treadmill for fish." The "treadmill" is an artificial river current with a dusty powder added to make the movement of the water visible with a laser. He then plopped a cylinder inside the tank to mimic the eddies produced by rocks in a river and used the data to create a computer model of the scene.

"If you put a cylinder in flowing water, what you see behind it is a staggered array of vortices. Little eddies," says Liao. "It's a repeatable pattern. And that pattern you can change slightly by changing the diameter of the cylinder and how fast the water is flowing. We can make the water go faster. We can make the vortices bigger. We can see what effect it has on the fish."[100]

If you're a fish swimming upstream and there is a rock in front of you, vortices on the left will peel off, then vortices on the right, then the left again—and they're all hitting you in the face because you're facing upstream.

"But the fish were doing something extra," says Liao. "There's an energy in the environment, and the fish is a kind of biological windmill. It's capturing this energy somehow."

This piqued Liao's curiosity, so he used what are basically fine-wire electrodes to measure the fish's muscle activity. "It's almost like acupuncture for fish," he says. The muscles of a fish swimming in calm waters fire in a wavelike pattern from head to tail. This is also true for a trout in a river with no rocks. Once a rock is placed in the water, however, almost all muscle activity stops, even as it "swims." Every once in a while, activity fires near the head, but that's it.

"This was surprising!" says Liao. "If you look at the video of the fish, the fish was bouncing around like crazy behind the cylinder and not drifting downstream at all. So that's what led me to this knee-jerk idea: 'Well, if a live fish isn't using any muscle activity, or very little, how different is that than having a dead fish towing behind a cylinder?'"

Liao tested this question using a recently dead fish that hadn't yet stiffened with rigor mortis. The dead fish moved upstream, got caught in the suction zone, hit its head on the cylinder, and fell back. It did this over and over again, maintaining its upright balance rather than flip-flopping madly about like a fish rigid with rigor mortis. The dead fish displayed a well-guarded secret: fish don't always need muscles to swim. "They are exploiting the energy of these vortices," says Liao. "It's similar to an eagle flowing on a thermal in the sky."

To swim upriver, trout don't resist the water; they go with the flow. They relax their fins and let their bodies flap side to side like a flag in the wind. The spinning eddies and turbulence propel them forward, turning a possible constraint into a blessing. The flexibility of fish spines allows them to slalom in between the vortices. Each eddy contains energy that causes a pressure drop, which the fish use to enhance their swimming performance, engaging only the muscles near their head to pilot themselves. The trout's body is perfected to exploit this rotational energy. If the timing is right, a dead trout truly can "swim" upstream. The team dubbed this movement the Kármán gait.

What happens if there are no rocks to create eddies in the stream? Here's the catch: trout make their own. They produce vortices by opening and closing their gills, spawning swirls of water along the tail end of their bodies. The pressure from the vortices helps push the trout upstream. To accelerate, the trout opens its gills wider, increasing the scores of swirls that counter thrust and give it a boost. Flicks of the tail also help create negative pressure behind its body. The trout react to the vortices using a sense organ called a "lateral line system," which is a series of tiny U-shaped tubes along the sides of their bodies. At the base of each U is a wiggling hair that sends nerve signals to the brain about where the vibrations are coming from. This nimble mastery of vortices is fundamental to their fast bursts and agility, the fish's flesh and blood proving far superior to the stiffness of humanity's ships and underwater robotics.

When Liao showed his father the videos of the fish slaloming between vortices, he said, "Oh, that's just like *wu wei*," a Taoism philosophy that means you do nothing and everything gets done. The saying is not an invitation to laziness but a state where you are at peace even when engaged in a frenetic task, completing it with skill, efficiency, and tranquility, just like a trout during its epic upstream migration.

"Trout have been in these streams for millions of years before us and will probably be for millions of years afterwards," says Liao. "They're just doing their thing, and we're trying to peek in and see a portion of what they're doing."

Capturing Energy From a Vortex

"MR. BERNITSAS, ANYBODY who has dealt with vortices has drowned."[101] That was the word of caution Professor Michael Bernitsas received as an undergraduate when he told his professor he wanted to take the plunge into vortex research. Forty years later, he's alive and well, and he's still got his eye on vortices, this time with a patent in hand. With the University of Michigan, Bernitsas has developed a trout-inspired technology called vortex induced vibration for aquatic clean energy (VIVACE).

As the name of his invention suggests, his aim is to capture the energy of vortices. Why? According to the United Nations, energy is central to nearly every major challenge the world faces today, including poverty, gender equality, adaptation to climate change, food security, health, education, sustainable cities, jobs, and transport. With more than 70 percent of Earth's surface covered in water, scientists are looking to tap its potential. One of our options is hydropower, which captures fast-running water to produce electricity. But there are economic and environmental pitfalls that make this a complicated option. The dams required to harness hydropower can prevent fish from traveling upstream, erode the nutrients of rivers, and hurt ecosystems.

Rain is another possibility, perhaps through hybrid solar panels that can collect energy from the sun as well as the energy of impact from raindrops splashing on the panel. If this were to materialize, rain would be a small addition to our basket of energy sources.

Then there's VIVACE, which makes use of this "vortex shedding" in currents to generate energy. Cylinders are plopped into a river perpendicular to the direction of the current, swirling the water flowing around them into whirlpools that alternate above and below the cylinders. The whirlpools bob the cylinders, moving a magnet up and down a metal coil, which in turn creates a DC current that is changed to AC and sent to shore. The machine then captures this energy.

The team's ambition is to exploit an underused potential of energy: slow water currents. Most turbines and open-water propellers need at least four knots for operation and five knots for financial viability. VIVACE is designed to function in currents as slow as two knots. Unlike hydroelectric plants, which typically require dams or penstocks, VIVACE doesn't significantly change the pathway of a water stream. Yet despite VIVACE's successes so far, vortex technology is one of those ideas that hasn't completely severed its umbilical cord from the laboratory. The industry has a small set of loyal followers who advocate for its use, including those who have graced the pages of magazines for their attempts to create an atmospheric vortex engine, essentially a controlled tornado on land, that produces clean energy. Many of these projects have garnered more media might than field use, primarily because it would cost hundreds of millions to scale up the technology. But that doesn't mean it will prove fruitless in the future.

If anything, the legacy of vortex research so far is in the potential of its power. A Kármán vortex street, or a series of whirlwind vortices, brought about the collapse of the famous Ferrybridge power stations in West Yorkshire, England, in the autumn of 1965, when eight newly built cooling towers spun vortices behind them, which slammed into three of the towers in the back and caused them to crumble. A thrill ride at an Ohio amusement park suffered a similar fate when vortex shedding caused one of its three towers to tumble over (luckily the park was closed for winter at the time). Nowadays, tall structures such as smokestacks benefit from strakes, or bumps, to deliberately introduce turbulence and break up a potential vortex shedding event. The research into vortices is not static but, like a river, ever changing—just as the creatures behind the inspirations are not final versions but will evolve in the future too, perhaps surprising us again.

"It is ironic," says Dr. Fish, "that an animal exploited close to extinction by humans should provide us with the inspiration to better our own future."

12

WINDOW PAIN

Spider Webs Inspire Bird-Safe Windows

"We have the dubious distinction of being the deadliest species in the annals of biology."

YUVAL HARARI, historian

A BIRD DANGLES upside down by its foot, caught in a finely woven forty-foot (twelve-meter) net, an hour's drive southwest of Pittsburgh in Rector, Pennsylvania. It's a crisp September morning at the Powdermill Avian Research Center (PARC), with blue skies peeking through creamy clouds. The preserve is surrounded by lush fields and cheeping birds with their feet in the air. A tufted titmouse with a gray mohawk and a peachy wash of color on its flanks is in a similar predicament. A cedar waxwing hangs like a confused watercolor painting, orange-tinted feathers fading into white and yellow.

A researcher carefully untangles a red-eyed vireo and stows the cream-bellied bird with olive-green wings in a cloth bag. Nearby, he bags another. Approximately 13,000 feathered fliers dangle each year from the center's nets and get their ankles banded. Some also go to "the tunnel," a dark wooden shaft the size of a small houseboat. At the far end of the thirty-three-foot (ten-meter) tunnel, light streams in from a window with two different glass panes, side by side—one a traditional pane of glass and another crisscrossed with ultraviolet (UV) lines. The question the researchers hope to answer here is: Do birds see in ultraviolet?

We certainly don't. Humans see only a thin slice of the electromagnetic spectrum—0.0035 percent of all the range of light that exists in the universe. We see wavelengths in the visible realm, with red being the longest at around 700 nanometers and violet the shortest at 380 nanometers. Anything above or below that range is invisible to our naked eyes. Color for us is not a reality "out there" but a collaborative creation with the light, the eyes, and the brain. Take the old question: If a tree falls in the forest, and no one is around to hear it, does it make a sound? Now ask a similar question about color: If there is an apple and no one is around to see it, is the apple still red? The answer is "Not really." Other creatures may not see a red apple at all. A dog sees a banana as yellow but the red apple next to it isn't red; it's dark brownish gray. A mouse sees only blue and green light, so the apple looks the same to the mouse as it does to humans with red-green color blindness.

In 1801, German chemist and physicist Johann Ritter wanted to explore the limitations of our sight further. He split sunlight using a prism and then measured the darkening of photographic paper for each wavelength of light. He found that the paper darkens more quickly when hit with shorter wavelengths like violet light, so he exposed the paper to light rays just beyond violet and found this produced the most darkening, proving the existence of another light we cannot see, which was eventually called "ultraviolet" light.

If we zoom inside our eye, we'll see why ultraviolet light is invisible to us. Our retina has three types of light detectors: blue, green, and red cones. Each cone is named based on its peak sensitivity to that color, but they also detect a range of other colors. The cones overlap and allow us to see more than just three colors. For example, the yellow banana triggers the green and red cones, but not the blue. When all cones are stimulated equally, the brain sees white. Many fish, reptiles, and birds have a fourth cone that dips into ultraviolet light.

A rare few people see beyond the ordinary scope of human vision. Patients with aphakia are missing the lens of one or both eyes (whether they were born that way, had them removed surgically, or lost them to injury), and since lenses help block ultraviolet light, the vision of these people sometimes dips into ultraviolet light, with postoperative patients, for example, reporting certain whites as now having a whitish-blue tint. Perhaps the most famous patient with aphakia is French impressionist Claude Monet (1840–1926), whose vision began to decline in his sixties due to age-related cataracts. "Colors no longer had the same intensity for me," he complained, "reds had begun to look muddy" and "my painting was getting more and more darkened."[102] To avoid mixing up his tubes of colored paints, he placed labels on them. Finally, at the age of eighty-two, he decided to undergo cataract surgery. When he recovered, he began destroying his prior canvases, distraught by what he'd rendered under the influence of his prior eyesight. Post-surgery, Monet's flowers resembled those in his artwork before his age-related cataracts, except this time the water lilies had a bluish hue in the white pigment, perhaps from the ultraviolet light he now saw reflecting off the flower petals.

But what about those birds? The question of whether they can see what we cannot is deceptively simple, but the stakes are high for conservationists and bird lovers around the world. An absurd number of birds die each year from window collisions, somewhere between 100 million and 1 billion in the United States alone.[103] This accounts for up to 5 percent of the total wild bird population—and that's just in one country.

Luke DeGroote, an avian research coordinator from Carnegie Museum of Natural History, is an investigator on the bird-collision case. A fair-skinned, fair-haired man with an easygoing demeanor, he sports a Powdermill cap as he shows me around the grounds. At the heart of the reserve is the bird-banding program; started in 1961, it's the longest continuously running program of its kind in the United States. In the last couple of decades, PARC has added window experiments to its research. DeGroote beelines for the tunnel, which is believed to be one of only two in the world. This impressive fact belies its plainness—a wood tunnel with a net inside, two types of glass slotted into the far window, and a side door to release the birds. Today, a woman named Rose is at the tunnel's helm, in blue jeans and a white cap with a neck flap. Four cloth bags droop from hooks hammered into the wood outside the tunnel. Rose grabs one, loosens the drawstring, and plunges her hand inside. She withdraws a small bird, its sage-colored wings folded snugly against her palm—a young red-eyed vireo. You can tell he's young, says Rose, because his brown eyes haven't yet morphed scarlet, giving him the red glare of human eyes caught in a camera flash.

Rose checks the bird's band number, written in minute script on an aluminum cuff around his leg, and says

"Red-eyed vireo 68534" to a video camera before releasing the bird through a fist-sized sleeve in the tunnel's wall. The bird flaps toward the light streaming through the window at the far end of the shaft. He gains speed, flying toward freedom, his eyes on the sky ahead, and then... thwaps into a net for the second time that day. Rose, satisfied with his performance, walks to a side door to free the bird. Not all birds take to the tunnel like this red-eyed vireo. Some hop around, confused, like a broken windup toy. Others don't move at all. The first red-eyed vireo Rose tried that morning just clung to the ceiling. "Vireos are not the most cooperative," DeGroote explains. "They usually hang out in the canopy. Put them in a dark tunnel, maybe they're like, 'I don't know what's happening, I'm just going to sit here.'"[104] Only some birds are tested in the tunnel because certain species risk injury based on how they fly. Hummingbirds, for example, are a no-go even though they're common window casualties. Robin-sized birds or bigger are skipped because they could plow right through the net. "A lot of sparrows, a lot of warblers, a lot of thrushes," says DeGroote.

That brings us to another glaring issue: windows are not equal-opportunity killers. The most common victims are migratory birds that fly hundreds, sometimes thousands of miles only to die by a mirage; they see bushy vegetation or empty skies reflected on a windowpane and try to glide through the illusion. But what they see as freedom is actually a death trap. Danger peaks during the Great Migration when, day or night, there is a bird somewhere in the world on a journey to a new land. Their pilgrimages are truly remarkable. The four-ounce arctic tern—all white bar a black kippah-like patch on its head—flies from one pole of the Earth to the other, a sojourn of more than 18,000 miles. Its journey is mostly overseas, with no buildings in sight, but for common species like hummingbirds, catbirds, and thrashers, the threat is real.

When I shared with a friend the number of birds that die from windows each year, she raised her eyebrows and asked, "So what?" Billions of dead birds *seemed* a good enough answer, yet the more I thought about it, the more I felt compelled to address why it does, in fact, matter. Only recently has this question entered the public consciousness. Prior to the 1800s, we were clueless about where birds disappeared to when they migrated. Birds were feathered riddles: Do they fly up and away? Do they mutate into other bird species, as Aristotle believed (the redstarts into robins, garden warblers into blackcaps)? Do they hibernate at the bottom of the ocean? Charles Morton, a Harvard vice-president, wrote a treatise that boldly claimed birds flew to the moon.

It wasn't until the 1820s that, in a prophetic twist on European folklore, a white stork helped to birth the migration revelation. A stork with a thirty-inch Central African spear running through its neck was found near the German village of Klütz. The flier had survived the attack in Africa and had then flown 2,000 miles with a spear protruding from its neck—only to be shot down again in Germany, this time fatally. And that's when the feathers started to add up. The stork, and later twenty-four others like it, helped pierce the heart of the migration quandary: if the spear was African, then the stork must have flown thousands of miles. Evidence had undeniably fallen from the skies, and just like that, the world shrank a bit—a stork connecting a hunter in Africa with a hunter in Germany. We now know some

4 billion birds migrate south from Canada into the U.S. each autumn, while another 4.7 billion leave the U.S. to fly over the southern border to the tropics.

Windows are especially risky business for first-year birds that haven't yet experienced the city's bright lights and glass. Many act as if they had beer goggles on, bashing into windows with clumsy boldness. Warblers and flycatchers take a big hit, possibly because they swoop through small spaces in forest canopies to catch insects and try to do the same with the reflections of trees cast on windowpanes. In a double whammy, the bugs they hunt are lured to artificial light, emboldened by a bright source of navigation. As nocturnal critters, moths travel by the glow of the moon, tuned to the faint radiance in the darkness. When a bright light appears, it is irresistible. Birds dive on the dazed insects and grab the little caloric snacks, but the longer birds dine in the realm of human edifices, the greater the jeopardy.

Most of us have, at one point or another, startled at a bird whacking into a window; despite feeling saddened, we go on with our lives. But those whacks add up, and collectively, they leave a dent on humankind. Birds consume up to 98 percent of certain insect pests, keeping their grazing propensities in check and helping to boost our agricultural production while also cutting down on our need for pesticides. Estimates indicate that birds eat some 400 million metric tons of flies, ants, beetles, moths, aphids, grasshoppers, crickets, and other arthropods around the world each year. Without their collective hunger, crop destruction from bugs would run amok. "The global population of insectivorous birds annually consumes as much energy as a megacity the size of New York," says Martin Nyffeler of the University of Basel. "They get this energy by capturing billions of potentially harmful herbivorous insects and other arthropods."[105]

On more than one occasion in my garden, I've witnessed an aphid strutting on a leaf, camouflaged in green. With vegetables, the largest of troubles always seem to stem from the smallest of trivialities, and aphids are no exception; they are the enemies to gardeners all over the world, piercing leaf vessels and sucking sap with vampiric ease. Moreover, the aphid's parthenogenesis prowess means some aphid females need a male like a fish needs a bicycle; they can deliver five green babies a day without the need for sperm fertilization. The babies mature within a week and can push out the same number of offspring. Fortunately for gardeners, many birds such as hummingbirds and American sparrows take delight in swallowing the little critters. So what is being done to protect our planet's birds? A generation ago, we seemed to think that a hawk sticker would do the trick, but that isn't so. A single hawk decal leaves a lot of open window space that isn't marked. Many more stickers would have to be placed on the window for them to be effective.

Seeing a bird-friendly need in the window industry, German company Arnold Glas swooped in to fill it. The company's vision stems from a paper written in the nineties suggesting that spider silk reflects ultraviolet light to prevent birds from busting through webs in the wild, allowing spiders to preserve the fabric of their home and their ability to capture prey. (This finding has since been debated. The ultraviolet in spider silk could also be used to lure insects to the web, to help the sticky strands resist damaging rays from the sun, or a combination of all three.)

The remarkable properties of spider silk have long been recognized. The ancient Greeks used the silk to dress wounds and stop bleeding, and astronomers in the nineteenth century used the fine lines as crosshairs on their telescopes to track the stars. Inspired by the simplicity of the spider's UV design, Arnold Glas made glass windows with a weblike coating that reflects ultraviolet light. The pattern resembles a game of pick-up sticks, where the sticks are strewn across the table and crisscross in a random pattern. The windows were "the only UV bird-protection glass in the entire world that was on the market around 2010," says Lisa Welch, who has been working with Arnold Glas for about a decade. "My early roles were not about marketing the glass, but market development, because the bird-friendly glass market was basically nonexistent back then."[106]

Now, demand is rising fast. UV windows have been installed on the Chase Center in San Francisco, the S. J. Quinney College of Law at the University of Utah, and the Building Society Gartenstadt Wandsbek office in Hamburg, Germany, to name just a few. Recently, New York City passed landmark legislation, with a vote of forty-three to three, requiring all new construction and major building renovations after January 10, 2021, to use materials that prevent bird fatalities, such as fritted glass (which uses ceramic-enamel coatings in a pattern like dots), tints, or UV-glazed windows.

But we still haven't answered the core question: Do birds *actually* see the ultraviolet lines hidden in the glass?

A Room With a View

"WE ARE PRISONERS of our own senses," says Graham Martin, an ornithologist and emeritus professor of avian sensory science at the University of Birmingham.[107]

A gentle-spoken man with white hair and glasses, Martin has studied bird senses for a lifetime. He's published papers on more than sixty bird species over the years, from spoonbills to penguins, but he has had a soft spot for owls and oilbirds, the latter of which are brown with white speckles and spend most of their lives in caves, navigating the darkness with echolocation. Oilbirds, locally known as the *guácharo* in the northern regions of South America, are so named because they feed on the region's oily palm and avocado fruits. It's perhaps ironic that Martin and I discuss the world of avian senses through video chat, which inherently inhibits our own sensory experience. I see only Martin's head and shoulders in a navy plaid shirt and jacket, the bright glare of light off his glasses obscuring his eyes.

I ask for his professional opinion about ultraviolet windows. The purpose behind the innovation is to inscribe the windows with something like avian hieroglyphics that only birds can see, warning them of danger; the UV says, "Humans have made something see-through; do not be fooled, my friend." But do the birds see these lines?

Well, first things first, says Martin: "There is no such thing as bird vision." That would be like saying there's such a thing as mammal vision. "Different species use different information at different times and we've got to take into account the diversity." A pigeon with eyes on the sides of their head, he explains, has a three-hundred-degree visual field (humans have 120 degrees). "We know nothing of what's going on over here," says Martin, waving his hands behind his ears. "I can't see my hands right now. If I were a pigeon, I'd see everything. If I were a pigeon and I wanted to 'talk' to someone, it'd be sideways because my vision is best that way."

Similarly, there is no single sensory reality in the animal kingdom. We have only to think of an owl at nightfall, snatching a brown mouse from a field of tangled shrubs, to feel the certainty of this truth. Even the eye of a fly, arguably the lowest creature in the gaze of humanity, absorbs a deluge of three hundred frames per second with its 3,000 lenses, whereas we gather a paltry sixty frames per second in bright light and just twenty-four frames per second in low light. It's as if all of us creatures have been given different keys to unlock the mysteries of the world around us.

"Ultraviolet seems like quite a good idea," says Martin. "Everybody says, 'Oh yeah, passerines have got UV vision,' but they're actually not that sensitive to UV." These birds hit the sweet spot somewhere in the middle of the spectrum—the greens and variations of that. "That's the most sensitive, and you drop off to the reds and the UV. They do see something in those wavelengths but they're not that sensitive. If you're relying upon just UV-reflective patterns to trigger them, you've got to have a lot of UV light to make it conspicuous."

Window experiments performed in the PARC tunnel are trying to test just that, coming up with threat scores for what the birds see. For example, the threat score for UV windows is around twenty, which means that for every one hundred birds they test, eighty could tell the difference between UV windows and plain glass, while twenty could not. This may not sound like much, but anything with a threat score of thirty or less is considered effective. A score of twenty or less is considered highly effective. But reality is also messier and more chaotic than experiments in the tunnel. UV windows are muddled by bright indoor lights and reflections from bushes and trees. Tests like those performed in the tunnel are a good start, but they're not 100 percent representative of what birds see in cities (although they are necessary for preliminary data). Bright days also turn to rain, snow, or overcast weather, and this makes it hard to predict bird behavior; UV glass, when it works well, only truly does so in bright light. Songbirds, in particular, are hit with high fatalities, likely because they migrate at night, fly low to the ground, and tend to be disoriented by artificial light. Other common deaths include the ruby-throated hummingbird and the yellow-bellied sapsucker, a stout woodpecker with a bright red forehead. But there's almost a hush about UV windows not being a clear win: if you don't stand with ultraviolet windows, you stand against birds. Is there any other option?

Turns out there is. Back at PARC, Luke DeGroote trots off in the direction of his office at the nature reserve. "I'd say parachute cords hung four inches apart are the most effective," he says, fiddling with strands on a device called the Acopian BirdSaver. "They sway a little. When we tested them in the tunnel, the threat score was like five." Nearby, he shows me UV-reflective chart tape developed by the American Bird Conservancy. "The chart tape is pretty close," he adds, referring to strips of tape applied to the outside of the window to disrupt the reflection.

If cords and chart tape are low cost, easy to install, and top tier in performance, why are we not using them more?

They work, everyone says, but no one except avian researchers will use them. Rather than asking uncomfortable questions about our need for an unobstructed view, we defer to an option that is second-rate but makes us feel good; "We did something," we can tell ourselves, with some relief, at the end of the day. Knowing all this, you'd think the legislation would be complicated, but it's remarkably simple, all things considered. Cities such as New York, Toronto, and San Francisco now include bird-safe guidelines for buildings. Local Law 15/20 in New York City specifies that 90 percent of the first seventy-five feet of a building must use bird-friendly windows. In an interesting twist, New York City has "both the strictest of standards (coverage areas, building/project types, no-exemption relative to geography) and also the simplest to apply in a broad sense," writes Heath Waldorf, a principal consultant at Bird Control Advisory in New York. There are "no issues with reflectivity of glass specifically (Toronto), no determining proximity to urban wildlife areas (San Francisco), and no provisions requiring particular lighting design and controls (all of the previous)."[108]

UV windows are also expensive, and not all experts concur it is worth the price. Collision rates depend on a medley of factors, and the good news is that not every window needs a bird deterrent: Do the windows reflect bushes and trees outside? Are they curved? Angled? Parallel? How high? What season is it? What time of day? UV works to some degree, but the research is nascent on what glass layer the UV needs to be in for the birds to see it well. We know that UV on the surface layer—the outside of the glass—performs best because it reflects light naturally. But what about the other layers?

"Some companies seem to be better at doing it than others," adds DeGroote, who says he'd like to perform more tests.

There is one agreed-upon solution: cheap, well-placed decals or cords outside the glass. Keep up the good fight in searching for ways to make windows transparent and bird friendly; in the meantime, step outside if you want a clear view of nature. Easier said than done. You'll be hard-pressed to see window installations with cords dangling on the outside; it's just not aesthetically pleasing. We are more than prisoners of our senses; we're bound by our desires too.

As I wrap up my time at the reserve, I ask DeGroote what his favorite bird is. He's quick to reply: the blackpoll warbler. "It's only about this big," he says, gesturing with his thumb and forefinger about five inches. "I think they look really nice with their black cap and black stripes. It's sort of tuxedo-y." Their high-pitched staccato songs float through the forests of Canada and they get "squishy" to prepare for their migrations. "They have so much fat on them," says DeGroote. "They can fly from New England to South America in one shot. Seventy-two hours they just go." They weigh only half an ounce and yet fly 1,800 miles over the Atlantic Ocean, one of the longest flights for a migratory songbird. In fall, they molt and transform, their plumage mutating into a dusty yellow green.

"It took me four months to walk the Pacific Crest Trail [2,650 miles]," says DeGroote. "These guys can go from New England to South America [4,200 miles] in three days." Their numbers are in swift decline, around 5 percent per year, according to the USGS Breeding Bird Survey. Habitat degradation and window collisions are the prime killers. How willing are we to sacrifice a clear view to preserve their numbers?

As I gaze out a window, my mind a whirling dervish of thoughts, a hummingbird hovers over a snow-frosted bush, its wings beating so fast they blur into stillness. The hummingbird zips away—away, I ponder, from possible death. Should we sacrifice a clear view and switch to dots on our windows? Or should we preserve our view with UV and save some but perhaps not all bird species? What would you choose?

World Wide Spider Web

UV WINDOWS MAY not be a smash hit yet, but that doesn't mean there aren't other potential inspirations to be found in the arachnid kingdom. With more that 45,000 spider species on Earth, we certainly haven't exhausted the catalogs yet. There are spiders that eat bats, peacock spiders that shimmy in colorful mating rituals, spiders that eat flies with a side of pollen, spiders that make mini air bubbles underwater, and the Brazilian wandering spider's bite, which can stimulate an erection in men that lasts for hours. Spider silk, in particular, has garnered tremendous interest over the last decade. Some silks are five times stronger than Kevlar, while others are gluey liquids in humid climates; some enable electrical conduction, while others are temperature resilient or inhibit bacterial growth so mold doesn't grow on their lifelines. With so many options, it's no wonder researchers have found plenty to study here.

Take spider silk woven in a wet, humid environment that sticks to surfaces. When MIT engineers in 2019 were trying to make a double-sided tape that could join two slippery surfaces inside the human body, they looked to spiders that clear a dry patch from a wet branch using charged polysaccharides, swiftly sticking their silk lines to the now dry object. The team mimicked this design by creating a tape that absorbs wetness and then binds weak hydrogen bonds to the dry surface until a stronger group of covalent bonds (the sharing of electron pairs between two atoms) can form. The tape can suture fragile tissues, like lung and intestine, in just five seconds. With more than 230 million major surgeries each year, the team hopes to one day replace surgical sutures, which they call "a thousands-of-years-old wound closure technology without too much innovation."[109]

Spider genomes are large and notoriously difficult to sequence, which is why the bulk of arachnid research is still stuck in the lab. Most of our attention has centered on their silk, but Harvard scientists recently invented a compact depth sensor for micro-robots, wearable devices, and augmented reality headsets that were inspired by the eyes of jumping spiders. These tiny, hairy asterisks can pounce on prey several body lengths away with astounding precision despite their limited brain hardware. If we take a look inside phone and video game console sensors, we'll find that they integrate light sources and cameras to measure depth. The Face ID feature in smartphones, for example, creates a depth map of your face using thousands of invisible laser dots. This works for the hardware of a phone, but what about smaller devices like watches or micro-robots?

"Evolution has produced vision systems that are highly specialized and efficient, delivering depth-perception capabilities that often surpass those of existing artificial depth sensors," writes the team.[110]

Why didn't they look to human vision as inspiration? Humans use stereovision, meaning each of our two eyes collects a slightly different image. Our brain takes those two images, checks the differences, and then calculates the depth. To see this for yourself, hold your finger in front of your face and alternate closing each eye. Do you see your finger moving slightly? It's computationally burdensome for the brain to stitch together a scene from those two images. Humans have brains big enough to handle the processing, but jumping spiders have brains that could fit on the head of a pin. Instead, jumping spiders have evolved two principal eyes, as well as two small lateral eyes, with semitransparent retinas arranged in layers that measure multiple images with different amounts of blur. What does this mean?

Let's pretend a jumping spider is looking at a fruit fly: the fly will appear blurrier in one retina's image and sharper in another. This difference in blur encodes information, providing the spider with data about the distance to the fly. This idea is already used in computer vision, but it requires big cameras with motorized internal components. To overcome this obstacle, the Harvard team created a "metalens" capable of producing two images with different blur. An algorithm interprets the two images and builds a depth-perception map. This allows for the team's depth sensor to be less bulky than existing designs, broadening the range of possibilities for its future uses in science and technology.

Of course, as many of us experience on a day-to-day basis, there is no such thing as a singular reality; our world is defined by how our brain perceives it—all that we see, hear, smell, taste, and touch is our way of navigating the world. Take an ant detecting movement through two inches of earth. If we did the same, it would be the equivalent of noticing a subtle movement thirty feet beneath our feet. Star-nosed moles can sniff out prey in streams by blowing bubbles into the water and re-inhaling them. Human vision pales in comparison to the panoramic view of chameleons, which can move each eye independently of the other and see in two directions at the same time. Humans may have large brains, but we still have blind spots in our vision (though few of us notice these visual holes). Like candles flickering in a mirrored maze, creating the illusion of endless candlelit halls, the external world and the workings of our brain blend seamlessly into a reality. Maybe part of truly seeing is learning to unsee, to ask ourselves: What am I missing?

13

FLASHES OF BRILLIANCE

Jellyfish Glow Inspires Nobel Prize–Winning Tool to Peer Inside Our Bodies

"Humans have learned various scientifically important matters, including genetics, by investigation of the various occurrences and mechanisms in nature; namely, we have learned from nature."

OSAMU SHIMOMURA, winner of the 2008 Nobel Prize in Chemistry

IN A MILLIONTH of a second, a tsunami of white light flashes and a ball of plasma hotter than the sun explodes across more than a mile of sky. Everything in the bomb's path is vaporized, including people, soil, and water. A mushroom cloud from the remains shoots high in the sky and supercompressed air expands outward, shedding its energy at the speed of sound and flattening houses, hospitals, and schools as if they were made of cards.

"A powerful flash of light hit us through the small windows. We were blinded and unable to see anything for about thirty seconds. Then, maybe forty seconds after the flash, we heard a loud sound and felt a sudden change of air pressure," writes chemist Osamu Shimomura.[111]

Born in Japan in 1928, Shimomura's formative years were shaped by World War II. When silver B-29 aircraft flew overhead, he would lie in a sweet-potato field and stare up at the mechanical birds. Other times, when an air-raid siren blared a warning, he hid in a ditch. On the day the bomb was dropped, Shimomura, then a teenager, walked home from his factory job in a drizzle of black rain, his white shirt turning deeper shades of gray.

Sixty-four years later, this boy would go on to win the 2008 Nobel Prize in Chemistry, along with Martin Chalfie and Roger Tsien, for contributions to "one of the most important tools used in contemporary bioscience."[112] The trio found a way to make glowing proteins shine a light on previously invisible processes in the body, such as nerve cell development in the brain or the spread of cancer cells.[113]

Three men won the prize, but many more helped bring the discovery to light. Molecular biologist Douglas Prasher was a hair's width away from earning his place in Nobel fame, while Ghia Euskirchen—a new graduate student in Chalfie's lab—inadvertently helped unstick two men's self-confessed "mutual ignorance" and get the work going again.[114] The story, says Shimomura in his Nobel lecture, "begins in 1945, the year the city of Nagasaki was destroyed by an atomic bomb and World War II ended."[115] Educational opportunities were limited in postwar Japan, so he found a spot at Nagasaki Pharmacy College, despite his lack of interest in pharmacy. Little did he know that that serendipitous choice would set him on a path to a storied career in chemistry. After a few years as a chemistry teaching assistant at Nagasaki University, he took a leave of absence to attend Nagoya University and broaden his scientific knowledge. His new mentor, Professor Hirata, assigned him to a project that had stymied esteemed scientists for decades.

And so it was that the twenty-seven-year-old visiting researcher found himself tasked with investigating how nocturnal sea fireflies (*Cypridina*) eject luminous blue light in the sands off the coast of Japan. Their emission of light—or "bioluminescence"—does not burn them from within like a bulb's heat. It is a "cold light," the energy from a chemical reaction rather than an electrical one. During World War II, the Japanese military used the sea fireflies as low-light lanterns on pitch-black nights as they trekked through dense jungles in the South Pacific. Torches or flashlights could alert the enemy to their position, a potentially fatal error. Discretion was key, but it was also nearly impossible to see your comrades on moonless marches. So they dried sea fireflies, pushed them into vials, and carried the creatures with them. At night, the soldiers twisted the caps off, crushed the sea fireflies in their hands, and smeared the bioluminescence on the backs of the soldiers in front of them. The faint blue glow allowed them to keep track of their comrades even as they crept fifteen feet apart to conceal their movements.

After the war, unused vials of the dried sea fireflies were shipped to scientific institutions for study. The glow had proved useful once; perhaps they could harness the light for future needs without relying on drying the creatures. Professor Hirata's lab received large shipments of the fireflies, and so, in the spring of 1955, Shimomura began trying to purify and crystallize the glowing compound—in other words, he needed to remove its chemical impurities to get solid crystals of the stuff. The substance was called luciferin—from the Latin *lucifer*, meaning "light-bearer." (In biblical texts, Lucifer is the devil prior to his fall from heaven, but in classical mythology it's the name of the planet Venus, personified as a male figure bearing a torch.)

Almost nothing was known about luciferin, so crystallizing the compound was a monumental task. It would have been easier if luciferin were a stable molecule, but it's not; it swiftly degrades when exposed to oxygen, like a fallen apple decomposing in sun-drenched soil. To ensure no oxygen ruined his sample, Shimomura performed the purification step in a hydrogen environment—a potentially perilous decision given hydrogen is highly flammable and can explode even at low concentrations. Shimomura began with 500 grams of dried *Cypridina* (about 5.5 pounds before drying), and after five days of round-the-clock work, he had purified just 2 milligrams of luciferin (the same weight as a small fly). He strived to crystallize the substance so he could understand its chemical structure, but all of his efforts "ended up with amorphous precipitates, and any leftover luciferin became useless by oxidation by the next morning. So I had to repeat the extraction and purification again and again. I worked very hard, and tried every method of crystallization that I could think of, without success."[116]

One day, after ten months of trial and error, he added hydrochloric acid to the luciferin. No ovens were available, so he abandoned the test tube on a lab bench to heat it another day. The next morning, his solution was colorless. He grabbed a microscope to get a better look at what had happened. Needlelike crystals came into sharp view. Success at last! It was a serendipitous discovery, and yet he'd accomplished something no one before him had. Giddy with excitement, Shimomura couldn't sleep for days. His success gave him hope for the future, which had seemed bleak since the war. From his purified compound, he was able to determine luciferin's chemical structure and oxidation products. In the spring of 1959, Shimomura's work led to an invitation: Professor Frank H. Johnson, a marine biologist at Princeton University, asked the young researcher to join his bioluminescence research lab. There aren't many photographs of Dr. Johnson, but in a rare portrait he is seen with round glasses and a white lab coat, his face glowing green from luminescent bacteria shining in a clear flask. Shimomura accepted.

In the summer of 1960, at thirty-two years old, Shimomura departed Yokohama aboard the Japanese ocean liner *Hikawa Maru*—a voyage destined to be the ship's last Pacific cruise. For thirty years, the ship had had an extraordinary life, from carrying Charlie Chaplin in 1932, to hiding Jewish refugees fleeing Nazi persecution in the early 1940s, to becoming a hospital ship during World War II. On the ship's last voyage, Shimomura was joined by hundreds of Fulbright fellows and students. Thirteen days later, he arrived in Seattle and boarded a Pullman car for a three-day train ride across the continent. Nearly a month after leaving Japan, he arrived in Princeton, New Jersey, and was met by Dr. Johnson on the platform.

Wasting no time, Dr. Johnson showed Shimomura his lab, and a number of vials of white powder. To most they looked innocuous, even boring. But, as Dr. Johnson explained, these were the freeze-dried light organs of a species of shimmering green jellyfish called crystal jellies, found in bountiful numbers in Friday Harbor, Washington State. The powder *should have* emitted light when mixed with water but, alas, it didn't. Would Shimomura like to study them? asked Dr. Johnson. Shimomura said he would be glad to help.

That summer (and every summer for the next twenty years during jellyfish blooms, it would turn out) Shimomura made the seven-day drive across the United States in a cramped station wagon with his wife, Dr. Johnson, a research assistant, and all his instruments and chemicals, their suitcases tied to the roof. They were headed to the San Juan Islands, an archipelago just off the coast of Washington where thousands of jellyfish float ghostlike in the waters, carried by the currents on a sojourn of the seas. Shimomura could peer into the clear waters and see crystal jellies flashing green like sparks from a wick. Friday Harbor was a true jellyfish mecca. There were other sorts of jellyfish too: the fried egg jellyfish with its yolk-yellow organs, and the lion's mane jellyfish with thousands of stinging tentacles that grow up to 120 feet long—a Rapunzel of the sea. To Shimomura, it was a paradise. But where divers would explore the waters by submerging themselves in its depths, Shimomura was an alchemist trying to harness the power of "cold light" by looking at how the jellies gleam in their watery kingdoms.

During their first summer at Friday Harbor, Shimomura and Johnson caught more than 10,000 crystal jellies to harvest the glowing liquid from their bells. To work with jellies today, thanks to new technologies, we only need to gather a few, isolate the bits we want, and clone them; it is faster and more sustainable than large-scale hunting. But back then, the work was repetitive: they dunked shallow nets to catch the jellyfish and filled bucket after bucket of the gelatinous jellies. With each jellyfish, they snipped off the brim of its bell, where the light-producing organs are, and squeezed the liquid out. Shimomura and Johnson worked long hours at the University of Washington laboratory trying to purify the jellies' bioluminescence, but they were unable to extract luciferin and luciferase from the rings of the jellyfish. On top of that, the two scientists had different visions for where their research should go next. Divided and frustrated, they worked separately in opposite corners of the lab. Nothing worked.

One day, needing space to think, Shimomura rowed a boat out into the harbor. Looking around, he wondered if the water's pH level of 8 was doing something to the jellyfish's proteins. He returned to the lab and found that a pH of 4 stopped the liquid from shining. When he added baking soda to bring the pH up to 7 (neutral), the solution glowed. But it was a faint light—nothing like the bright green the jellyfish create. Still, this raised his morale. If it was possible to inactivate the glow, then the solution could be reactivated—somehow—at a later point. He just didn't know the "somehow" part yet. Exhausted and with no answer in sight, he dumped his solution into a sink collecting overflowing tank water nearby. A flash of blue shocked his eyes! The sink lit up.

Had the tank's seawater sparked the bright flash? He investigated his hunch and, sure enough, calcium ions in the seawater had triggered the light. He reported the findings to Dr. Johnson, who was as excited as Shimomura. It was already known at the time that EDTA, a clawlike chemical that sticks to other substances, could bind to calcium ions. Their newfound understanding of the importance of calcium ions to the reaction allowed them to finally devise a method of extracting the luminescent material. In 1962, they used this technique to get 5 milligrams of nearly pure luminescent substance, which they named "aequorin" after the Latin name for the crystal jelly (*Aequorea victoria*). But when aequorin was activated with calcium, it shone bright blue—not green like the crystal jelly in the sea. The blue flash was as provocative as it was puzzling: Why did the aequorin light up in blue, not green?

Shimomura and Johnson were missing something.

Living Light

THE MYSTERY OF light has long beguiled mystics, poets, mariners, farmers, and scientists like Shimomura. Whoever and wherever you are, you have likely at some point spoken of this silent glory. Light has been consecrated by every religion and culture on Earth, and is one of the oldest symbols in existence. In the book of Genesis, there is the divine "Let there be light." In Hinduism, there is Diwali (the Festival of Lights), which celebrates light prevailing over darkness. The Buddha of Infinite Light guides us to wisdom and clarity. Jesus was said to be light ("I am the light of the world"), and Allah is the light of the heavens and the Earth.

In literature, light is a metaphor for all that is good. In Shakespeare's *Romeo and Juliet* it is a metaphor for beauty: "But soft! What light through yonder window breaks? It is the east, and Juliet is the sun." In common parlance it is the truth laid bare ("Let's shed light on this"), a luminous epiphany ("a light-bulb moment"), and hope after darkness ("light at the end of the tunnel"). Light is a symbol of purity, freedom, and progress.

But what is light separate from its symbolism? Over eons, light has transformed the face of the Earth, with millions of species growing eyes to capture light and plants soaking their leaves in the power of the sun's rays. Are heat and light inextricably linked? No, they are separate entities that can merge into a fiery glow. Their individuality is readily apparent too: hot water is felt but not seen, and the light of stars is seen but not felt. Is light made of particles? Waves? Something else? For the longest time, the question was a prodigious problem, one that took centuries of physics to come to terms with.

Light doesn't travel through a bent tube like sound does, reasoned Newton, therefore light cannot be a wave. Light, he said, is composed of particles. Light is a wave, reasoned English physicist Thomas Young, because two light waves can cancel each other out when a wavelength's trough meets the other's crest. Benjamin Franklin summed up his own confusion: "I must own I am much in the *dark* about *light*. I am not satisfied with the doctrine that supposes particles or matter called light are continually driven off from the sun's surface with a swiftness so prodigious."[117] Einstein came in with the tour de force idea that light is both particles and waves (later termed the wave-particle duality), and that the speed of light is constant, and time mutable. It is an astonishing thought, one that changes how we view the universe. The mutability of time stands against everything we build—our clocks, watches, and towering timepieces—and yet time on, say, the International Space Station is minutely slower than on Earth.

The seductiveness of light lies in part in our inability to produce it from within. Our bodies are solely sensors to this ethereal entity. Too little light and human bones soften and weaken in a disease called rickets that bows the legs. We call life itself light and use all sorts of metaphors for death being its absence: a candle extinguished; a person's light gone or taken from us; poet Dylan Thomas's "rage, rage against the dying of the light." We metaphorically put the light within ourselves even though it is "out there" and not within us. In envy or awe, we try to harness the power of light, and along the way, mishaps of Icarian proportions have befallen humanity: a tipped candle igniting a blazing blight across seventeenth-century London, smoke from kerosene lamps destroying lungs, and massive wildfires sparked from a glowing cigarette butt.

And yet, creatures in the ocean have pulsed with an inner glow for millions of years. Over the centuries we've recorded the phenomenon in awe. In the first century, the portly Roman philosopher Pliny the Elder observed creatures glowing near Mount Vesuvius, including a clam-like species of mollusk in the Bay of Naples that bores into soft rocks. Those who ate the creatures, he said, shone from their mouth: "It is the nature of these fish to shine in darkness with a bright light... and in proportion to their amount of moisture to glitter both in the mouth of persons masticating them and in their hands, and even on the floor and on their clothes when drops fall from them, making it clear beyond all doubt that their juice possesses a property that we should marvel." He also noticed that the slime of a jellyfish species rubbed on a walking stick "will light the way like a torch."[118] French philosopher René Descartes compared the sparks of light in agitated water, likely from bioluminescent plankton, to "pieces of flint when they are struck."[119] Natives of the West Indies tied glowing beetles between their toes as they walked in the dark forest and applied paste from crushed beetles to their hands and faces.[120] Missionary and mathematician Guy Tachard said the sea was filled "with an infinity of fiery and luminous spirits."[121] Coal miners trapped fireflies inside glass jars as biological lanterns to light their way. Off the coast of Point Reyes in Northern California, plankton glimmer blue at night when disturbed with the stroke of a kayak paddle.

A dazzling 90 percent of creatures in the deep sea are bioluminescent, usually with a blue-green light. Blue light penetrates the farthest in the ocean, giving the water its iconic indigo hues. Colors in the red, orange, and yellow wavelength are absorbed and thus removed at deeper depths. A scuba diver in a bright red diving suit appears all black fifty feet below the surface. There are, of course, exceptions; a rare few sea creatures have evolved in antithesis to the trend. The deep-sea black loosejaw (so named for its open-hanging jaw) uses red light to hunt for prey. Most sea creatures cannot perceive red light, so it acts as a camouflaged flashlight in the vast darkness. At about 3,300 feet (1,000 meters)—the no-light (aphotic) zone—sunlight is no longer detected, even with the most sensitive eyes. Of course, the no-light zone is a misnomer; it is a veritable light display down there, a party of creatures that flash light, no sun required. The male sea firefly zips around and squirts bright mucus to let females know his whereabouts. The deep-sea shrimp spews glowing goo to distract predators as it escapes. The anglerfish has what looks like a stick poking out of its head, except it's tipped with bioluminescent bacteria to lure prey to the fish's razor-sharp teeth.

Then there are the brainless, spineless jellyfish. These ancient seafarers have drifted in Earth's oceans for longer than any other multi-organ animal on the planet. As creatures go, jellyfish veer toward extraterrestrials with their astonishing diversity and unusual abilities. There are more than a thousand species, including those that clone themselves, those that lay 45,000 eggs in a single night, and those that can kill a person in minutes. Then there is the immortal jellyfish that, when aging or sick, returns to its infancy—a feat of reverse aging no other creature has mastered. Jellyfish do all of this while being 95 percent water, with the rest made up of proteins and minerals. As for the crystal jelly, *why* they shimmer remains a mystery to this day, but Shimomura wasn't interested in the *why*. He wanted to know *how* they shine without using energy from the sun.

Teamwork Takes the Prize

FOR YEARS, SHIMOMURA was stuck on the problem of what made his solution glow blue in the lab when the crystal jellies shine green in the sea. He suspected it had something to do with another fluorescent protein he found while trying to purify aequorin. But the creatures produced so little of this other protein that it took his team more than a decade to gather enough of what became known as "green fluorescent proteins," or GFPs, to study.

Luminescence, like that of aequorin, is not the same as fluorescence. Luminescence is the spontaneous emission of light and requires a chemical reaction (thus the calcium ions affecting aequorin). Fluorescence occurs when a molecule absorbs the energy from a nearby light source and emits it back at a lower energy. What Shimomura finally found was that GFP absorbs the light emitted by aequorin and converts it into lower wavelengths around the 510 nanometer range, which is green light. What this means is that the proteins fluoresce green when exposed to light in the blue (aequorin) to ultraviolet range. And this makes GFP perfect for imaging, because scientists can use a UV light to trigger its glow, no chemical reaction necessary.

Across the country from the San Juan Islands where Shimomura was making these discoveries, Martin Chalfie (the second Nobel winner) was a biology professor at Columbia University. He wasn't a fluorescence researcher. He was a worm guy, and he almost didn't even become a research scientist. As a child, Chalfie was interested in science but not in the way he saw in his colleagues around him. He read books about insects, plants, and dinosaurs and cut cartoons of animals into scrapbooks. He didn't enter academic competitions or almost blow up his childhood home in a chemistry experiment gone wrong. His interests were wide and varied, and he entered Harvard University not knowing what he wanted to do. He was an average student, with his worst grades in physics and chemistry.

After what he calls a "disastrous summer lab experience" in which all his experiments failed, he decided the life of a scientist wasn't for him.[122] He became a high school chemistry teacher and did other, odd jobs, like working as a traveling salesman for his parents' dress-manufacturing business and setting up summer rock concerts, before he decided to give laboratory research one more shot. This time, he was more successful, and that success gave him the confidence to apply to graduate school. His meandering route eventually landed him in neurobiology research, with a specimen he would study for decades to come: *Caenorhabditis elegans*, a long name for a tiny, transparent worm the size of a comma that contains some one thousand cells.

C. elegans is what scientists call a model organism because it is a simple animal with a digestive system, nervous system, and reproductive system, as well as a short life cycle, meaning it reproduces quickly and is convenient for gene research. This research often involves peering down the lens of a microscope and looking at little worms wriggling in a petri dish. The creature was a niche but up-and-coming topic among select scientists, with the *Worm Breeder's Gazette* updating them on each other's progress. Chalfie began working on gene expression in the worm's nervous system; specifically, exploring the genes involved in mechanical sensation, including touch, hearing, and balance—senses that were not well understood in the 1980s. But when Chalfie found a potentially interesting gene, he could only observe what was happening in the worms using dead tissues. This made understanding how they change during development tough, because he had to compare images taken from different dead animals. It was like taking screenshots of a movie at different moments and trying to piece together what's happened in between.

Around the same time, biochemist Douglas Prasher was one of the first people to recognize the potential of green fluorescent proteins as a tracer molecule inside the body. Proteins are extremely hard to see, and using something that visibly glowed would be a game changer. To truly understand what's happening inside our bodies, we need to map the very small; the body is not a single entity but many trillions of cells, and each cell can hold around 42 million protein molecules (tens of thousands with different functions)—that's a lot of tiny machinery to understand, and it raises a lot of questions. For example, where do the units of ribosomes come from and how do they assemble to translate genetic code into chains of amino acids? A human cell may have as many as 10 million ribosomes, with 60 percent of the cell's energy dedicated to making them.

In the late 1980s, Prasher was one of the few researchers working on the GFP's gene. He received a two-year grant from the American Cancer Society for the protein's potential in tracing cancer cells, but that gave him only enough time to isolate and sequence the gene. Where Shimomura had determined GFP's final chemical structure, Prasher found how the jellyfish express that protein structure. Think of it this way: where Shimomura had read the finished book, Prasher was looking for the gene that expressed how to write said book.

When Chalfie heard during a lunchtime seminar in 1989 of the green proteins Shimomura had purified, it bowled him over. Perhaps he could use the green proteins as a marker in his transparent worms. The worms are as clear as glass, so maybe he could use the green fluorescent proteins like a highlighter to track cell expression. After the seminar, he learned that his idea was similar to Prasher's, so he called him up. Prasher was working on sequencing the GFP gene in its entirety, but he wasn't finished yet. Chalfie asked Prasher to call him when he was done; he believed his transparent worms were the perfect animal model to insert the GFPs into. Prasher said he would. He finished sequencing the gene and published a paper in 1992. The green fluorescent protein, he found, has 238 amino acids linked together in a barrel shape, with amino acids 65, 66, and 67 responsible for absorbing UV light and emitting a green glow. But there was no fanfare for his work, and little support for his research, so he abandoned the project. And he forgot to call Chalfie.

That might have been the end of it, had it not been for a new graduate student doing a rotation in Chalfie's lab. Ghia Euskirchen had just finished her master's degree in chemical engineering working on fluorescence. Chalfie was searching for a project for her when he uncovered the latest research on fluorescent proteins. And that's when he learned that Prasher had managed to sequence the full green fluorescent protein. Chalfie called up Prasher again, who sent him the CDNA clone. At the time, several researchers, including Prasher, had already tried to observe GFPs inside living organisms, to no avail. Biologists were skeptical about whether it could actually work. They believed something intrinsic to the jellyfish must be responsible for activating the glow. Chalfie instructed Euskirchen to give it another go, using a recently developed polymerase chain reaction (PCR) technique to copy the gene.

It's hard to imagine, but back then this now-standard PCR technique wasn't as prevalent in laboratory biology. Many other resources we now take for granted didn't exist either: PDFs hadn't been invented yet, so scientists had to photocopy journal articles from bound volumes in the library. A photograph taken through a microscope (photomicroscopy) required film, so scientists had to wait a day for their photographs to be developed. Since Euskirchen was a first-year graduate student, her lab work was relegated to the evenings after class. The process of trying to express GFP in bacteria took time, meaning it was nearly midnight by the time she inserted the gene into bacteria, slid the stuff under a fluorescence microscope in a darkened room, and saw, for the first time, glowing bacterial cells with jellyfish DNA. She sat in the dark, stunned.

As it turns out, Prasher's traditional gene-copying technique had inadvertently added some base pairs before and after the fluorescent gene, and that little bit of extra DNA was enough to prevent the GFP from glowing. Prasher had been less than a paper's width away from making the same discovery that netted a Nobel Prize—but he'd followed techniques that were traditional at the time, not PCR. Despite Euskirchen's quick success in the lab, she wasn't interested in working with GFP and left after finishing her rotation.

Chalfie continued the research, working for a year to figure out how to express ("turn on") GFP in just six of the worm's touch cells. He placed the jellyfish's GFP behind a promoter in the worm's touch receptor neurons, a region of DNA that encourages (or promotes) the process by which instructions in our DNA are converted into the assembly of a protein. Essentially, it serves as a kind of switch to turn a gene on or off.

Chalfie then injected this gene construct into the gonads of adult worms, and when they reproduced, the offspring glowed green under ultraviolet light in their four touch cells. Now he was able to see when during the worm's development their touch cells turn on and gain their role as touch sensors. The article detailing his discovery made the front cover of *Science* magazine on February 11, 1994—with an image of a green-glowing roundworm.

Roger Tsien is the last man in the story. Tsien is a biochemist from the University of California San Diego who wanted to take Chalfie's and Shimomura's discoveries up a notch. If Chalfie took an unconventional path for a future Nobel Prize winner, Tsien was the exact opposite. As a child in a middle-class home in Livingston, New Jersey, he played with chemistry sets and borrowed chemistry books from his school library. In high school, he won first place in the Westinghouse Science Talent Search, considered the junior Nobel Prize competition, for his project titled "Bridge Orientation in Transition-Metal Thiocyanate Complexes." He received a scholarship to Harvard and began his studies in chemistry and physics. While there, he also nurtured an interest in music and began taking a music course to match every chemistry class he took. Tsien was talented but he also worked hard. He received a full scholarship to study at Cambridge University and stayed at the institution until 1981, when, at the age of twenty-nine, he accepted a position as an assistant professor at the University of California, Berkeley. He was a prolific researcher, already with a dozen papers under his belt and another soon to come that would be cited tens of thousands of times. In 1989, he relocated to the University of California San Diego, a campus nestled under blue skies with a light ocean breeze off the coast of the Pacific Ocean. Here he would stay for the next twenty-seven years.

Tsien's contribution to the legacy of GFP is expanding the palette of fluorescent proteins to include other colors that glowed longer and with higher intensity. In 1994, Tsien showed that oxygen is vital for GFP fluorescence and that point mutations in the gene could be used to alter the brightness and color of the protein. Using DNA technology, Tsien replaced certain amino acids in various parts of the GFP, making it emit light in other parts of the spectrum.

He gave the colors delectable names based on their emission profiles: the yellow ones are "mBanana," the orange "mTangerine," and the pinkish "mRaspberry," to name a few. Together, they became known as the "mFruits" (the *m* stands for "monomeric"). "Our work is often described as building and training molecular spies," said Tsien in an interview. "Molecules that will enter a cell or organism and report back to us what the conditions are, what's going on with the biochemistry, while the cell is still alive."[123]

The only hue Tsien couldn't produce was red, a highly sought-after color. Red light penetrates biological tissues easily, which is useful for studying cells inside the body. Instead, two Russian researchers, Mikhail Matz and Sergey Lukyanov, derived a reddish fluorescent protein from *Discosoma*, a scarlet mushroom corallimorph that resides in a closely related order to reef-building corals. The protein they found was modified by Tsien to be less heavy and more stable. This fluorescent color he called dTomato (clearly indicating which side he falls on in the great tomato fruit/vegetable debate).

A Guiding Light

AT LONG LAST, scientists were able to "paint" processes inside the body like never before, and record their interactions. GFPs have since become a guiding star for molecular biology. In 2002, during a speech Tsien gave on accepting the Dr. H. P. Heineken Prize for Biochemistry and Biophysics, he likened the complexity of proteins to "town dwellers, individual protein molecules in a cell are born, get modified or 'educated,' travel around, and cooperate or compete with each other for partnerships. Some proteins emigrate from the cell. A few have the job of killing other proteins. Eventually all the proteins will die at varying ages and their components will be recycled."[124]

Thanks to GFPs, scientists can now track which cell a particular protein calls home and follow its movements as it goes about its business. Researchers have used green fluorescent proteins to observe the migration of cancer cells, to learn how HIV spreads between immune cells, and to track arsenic pollution in our waters. The jellyfish's glow has even been used to detect explosive TNT. Estimates vary for how many land mines are buried worldwide, but they soar as high as 110 million. In 2019, around 5,500 people were killed or injured by land mines; of these, 80 percent were civilians, according to the International Campaign to Ban Landmines. Scientists have modified bacteria to glow green when they detect TNT or heavy metals like cadmium or zinc nearby. Bacteria are cheap, easy to spread over lots of ground, and take just hours to report back on the presence of land mines. The method was successful in experiments, but there are still hurdles that scientists need to overcome. For one, the sensors only work between 59°F and 99°F (15°C and 37°C). Two, even though this is done at night, the bacteria's light is faint in the field, especially during a full moon, so researchers are working on a device that shields the bacteria from moonlight. Drones will also need to be deployed so that humans don't get too close to land mines. Other uses for GFPs include tracking genetic inheritance, cellular growth, and protein interactions.

Green fluorescent proteins are even being used to probe one of the most mysterious of all our organs. The brain, a wrinkly three-pound organ protected in a helmet of bone, is composed of a labyrinthine network of interconnected nerve cells that communicate using chemical and electrical signals. Somehow, this maze of complexity gives rise to human consciousness. Of course, things can also go wrong in the brain, and they do. Alzheimer's disease litters the brain with plaque and clumps between neurons, disrupting their communication. At first, a person's memory slowly slips away, then mood swings and anxiety become more frequent. They struggle to remember new information, and words are harder to find. As their neurons and synaptic connections—shaped by time, environment, and experience—deteriorate, they no longer can take care of themselves and more memories slide into a fog. Their very identity is thieved from them. Scientists are now using fluorescent proteins to observe what happens to neurons in mice when plaque invades their connections. And yet there is still more to learn.

Scientists haven't perfected red fluorescent light, which penetrates deeper than green light in tissues like the brain. Most red fluorescent lights used today, including the red mushroom corallimorph, are a reddish orange. The best scarlet tags linger around 10 to 20 percent brightness, dim flashlights compared with the brilliance of green proteins. Could coral reefs harbor the elusive red fluorescent protein scientists have been seeking for decades?

It's certainly possible. Other animals offer hope, too—like copepods, shrimplike creatures related to the sea firefly *Cypridina* that Shimomura first studied so many years ago. Such a possibility has also become a heightened concern among scientists who are worried entire colonies of coral are suddenly losing their color and becoming ghostly white, nearing death. The most iconic, the Great Barrier Reef in Australia, is facing unprecedented devastation. Reefs in Guam, American Samoa, and Hawaii have experienced the worst bleaching ever documented. Around 98 percent of some reefs in the Northern Line Islands in the South Pacific have died, and more than 75 percent of all tropical reefs experienced heat stress between 2014 and 2017, of which 30 percent reached mortality. Are we losing other potential inspirations? It's hard to know. When GFPs were first discovered, Shimomura did not foresee how ubiquitous they would become.

The discovery is a shining example of basic research, a field that is increasingly hard to fund. If there is no immediate foreseeable benefit to humanity, basic research is unlikely to find a home in the lab. However, proponents of basic research argue that discoveries merge and connect to each other, like a vast network similar to the brain's own. It is a shifting mosaic, a chaotic choreography, with plenty of unknowns and missing pieces. Basic research is a study of the world and its fundamentals, but it is increasingly being seen as a secondary study to "higher" levels of research that have a direct, obvious impact on humans.

But what is so basic about light? The simplicity of the word belies its "everything-ness" in our world. The solar-powered plants we eat give us life. The light of the sun strengthens our bones. We evolved eyes that see the contour of mountains, rainbows after a rain, and the faces of our family, key actors in the theater of our minds. Without basic research, many formative discoveries wouldn't have been made. Take, for example, microbiologist Thomas Brock, who in the 1960s simply wanted to know what weird critters might be living in the boiling hot springs of Yellowstone National Park. When he looked, he found a species of yellow bacteria called *Thermus aquaticus* that thrive at 158°F (70°C) or higher. The bacteria's enzymes are stable at high temperatures—particularly Taq polymerase, which can replicate its own DNA. The discovery eventually led to widespread practical applications, such as PCR technology for the identification of DNA at a crime scene and, more recently, coronavirus testing. PCR technology earned biochemist Kary B. Mullis the 1993 Nobel Prize in Chemistry. Other developments that came out of basic research include X-rays and penicillin, both of which were being explored with no endgame in mind. Then there is Professor Emeritus Dan Kleppner, who told MIT that he wasn't "dreaming of developing the GPS" when he helped design the hydrogen maser, an atomic clock now used in satellite-based global positioning systems. "With basic research, you don't begin to recognize the applications until the discoveries are in hand."[125]

Green fluorescent proteins remain an incredible discovery; they have given us the ability to use nature's light to illuminate the secrets of our bodies. Some of the darkest places in this world are not "out there" but within ourselves, hidden inside the territory of cells, each with their own functions and finite lifetimes. How is it that all these diverse pieces inside us form something that feels as singular as a human body? Is this the greatest illusion of all time? The belief that we are singular in any way? Basic science "continually connects back to the world in unforeseen paths," Shimomura writes in his book *Luminous Pursuit*. "Thus it would be best that we remember to study nature in order to obtain new scientific knowledge."[126]

If astronauts, for example, were born in space and only able to look down on Earth, they would see a planet with swirling clouds, land masses, and vast oceans, but they would not see the fuss of activity of billions of people below. They would have no sense of the movement of life on this planet—from its ants to its elephants—its loves and its losses; the intricacy would be gone and only the largest, most noticeable patterns would remain. The astronauts would have to devise ways to gaze closer into the hub of activity on Earth, like scientists peering into a microscope at microbial activity in a petri dish.

Even now, the answer to "What is light?" would depend on who you asked. For the mystics it is a supernatural force; for the romantics it's a lustrous mystery; for the physicists it's a wave of electromagnetic radiation composed of massless tiny particles; for mariners it is guiding light; and for farmers it is the sustenance that feeds their crops. For the creatures of the deep sea, who know nothing of the sun, light comes from within to attract the attention of mates, to lure prey into their stomachs, or to send flashes of Morse code to each other. For medical practitioners laser light is a powerful healing tool, and for astronomers it offers clues to the cosmos. For geneticists it is a means to probe deep biological mysteries. Light is a virtuoso, a magician, a trickster that hasn't yet revealed all of its secrets.

A literature search these days for papers on "green fluorescent proteins" brings up more than a million results. All of this began with a man whose early life was marked by a flash of hot white light and the smoke of misfortune. Crystal jellies are now a vanishing species in the San Juan Islands where Shimomura spent his summers. An investigation found that experiments did not cause the decline, though what did remains a mystery. Guesses so far include shipping lanes, water contamination, and buildings near the shorefront. Now the question is: What do we stand to lose if they are gone?

Conclusion

IN SHARING THESE nature tales with you I have, out of a need for brevity, discarded many others that are just as wild and inspiring. I could fill another book with the stories I've uncovered, often on the recommendations of scientists I interviewed. These researchers flirt with nature, peer under the water's surface, and trace the veins of ancient leaves; they teach us just how connected we are to this Earth.

Nature is our best example of sustainability at hand. The living world, after all, has solved many of the dilemmas we face today with energy, food, transportation, temperature control, and packaging (fruit peels, seed shells, etc.). And yet for centuries, humanity has chosen to roost atop the animal kingdom and cast our gaze down on "inferior" beings. We patronized animals for their flaws, but we faltered in measuring them against man; they do not abide by our measuring sticks. Usain Bolt, one of the fastest humans on Earth, is outrun by a warthog. A warm-blooded tuna can dive deeper than a human and a micro-animal called a water bear can withstand temperature extremes better than us. Seen through unkind eyes, humans are feeble anatomical feats. We get impacted teeth, eat ourselves to death, develop blurry eyes, and are plagued with UTIs, and our skin ruptures with acne—it doesn't look good from the outside.

And yet, this evaluation is a simplification, a surface-level inspection of something more complicated. We can alter the very molecules of life. We can control the evolution of plants and animals with selective breeding. We can sequence our own genome and see inside our skulls. We can magnify our view of explosions in the cosmos and we can peer into the nature of our blood. We can replace failing organs and save those who are dying. We can capture energy and store it outside our bodies. And yet all of us, every living thing on Earth, are still caught in the same sticky web of survival and terminable energy. The world is truly a wild place constrained by shaky human hands.

Many inventions may seem to miss the point of biomimicry in terms of regenerative designs, but this is no surprise. We've lived with our technological expertise at this scale for only a couple of hundred years at best. It's a big request and responsibility to be a part of Earth in such a conscientious way. Rather than reprimanding ourselves for the past, we could be moving forward. The thought of losing one species is devastating enough, but the thought of losing that species and everything it might yet teach us takes it to a different level. We are speeding toward a turning point, our inordinate talent for making tools leading us—if we're not careful—to continued heaps of waste, depleted resources, species extinctions, degradation, and, on the whole, a less inspirational world.

In the twenty-first century, around half of the Earth's population lives in cities, connecting with plants only in their homes, in urban parks, or when walking past trees squeezed between pavement slabs and gum grit. East Asia's airborne emissions have ended up in central Oregon, and plastics from Tennessee have traveled to the Gulf of Mexico. The romance of a message in a bottle floating to new lands is not an ode to love but a stark reminder that borders are invisible lines drawn by humans and nature does not abide by them. We can draw lines in the sand and say *This side is mine*, but there are still microplastics in our salts, seasoning our carelessness. Life, as bewitching as it is, is not forgiving; it will tick-tock to destruction if we let it.

Rebuilding a better world may take slowing down and recognizing that speed at the cost of atmospheric destruction is not the price we want to pay. There is rebellion in choosing to turn from the speed of our hectic lives and refocus our eyes. To recognize with humble curiosity the myriad creatures around us, all bound by an equalizing bond: the unstoppable momentum of being finite on this Earth. After all, laying the foundation for a better future isn't just about the birth of an idea or place; it's about choosing where we want to go. Innovation simply for the sake of it is arguably a waste. If we idolize innovation, why do we place no value on what happens to our objects once they have served their purpose? Rarely do we consider an "end-of-life plan" for what we create. Creation is just as powerful as destruction, and yet we place little emphasis on the fate of our objects after their life of service. We simply throw everything away. Small garbage piles are then shipped off to gargantuan trash heaps. Our species is still young, trying to find its footing in a chaotic world, but we need to grow and acknowledge that this youthful stage is unsustainable. It is asking the hard questions to create a lasting, or at least more meaningful, existence. Of course, nature and human innovations differ in significant ways too. Humanity uses rotary motion for all sorts of machines, from propellers to wheels, but nature has made little use of this idea; wheels are the proverbial archetype of human invention. Mother Nature prefers legs, wings, and fins. Why not the wheel?

There are, in fact, several evolutionary obstacles that make the development of a wheel challenging. A complex structure like the wheel may be advantageous, but the intermediary structure required to get there provides no benefit and is even disadvantageous. Sudden change is only an option for engineers and inventors. Not everything about nature's creations is necessarily beneficial, either. Nature reinvents at a glacial speed, taking thousands to millions of years for dramatic changes to occur. The diversity of strategies life has used to survive certainly haven't enabled *all* species to thrive; however, life still exists on this planet after billions of years and multiple catastrophes. Perhaps, in some way, there is a smidge of hope in that.

The deeper we explore Earth's vast library, the more we learn; it is an adventure into uncharted territory. What if instead of conquerors, we see ourselves as guardians of this library? View the planet's creatures as portals of sorts—their lives, singularities, and otherness as beautiful as any magic we can dream up. What a delight that we get to learn their stories. What a gift.

I began this book with a reflection of mine, and it is with this I'd like to end: Although many of us may not leave riches to our families, we can bequeath them a legacy more precious than gold, more delicate than glass, and more monumental than fame: a world preserved in the amber of our protection.

Thank you for going on this journey with me.

Acknowledgments

THE BEST THING about being a writer is the ability to learn from an intelligent, adventurous, motley crew of people about their passions. Dozens of scientists provided space in their schedule to chat with me, often during trying times (like the COVID-19 pandemic and unprecedented wildfires), to discuss topics like fog harvesting, explosions in the cosmos, and species that have outlived all five major extinctions.

Though not everyone I spoke with is quoted in the book, they still provided valuable knowledge and were patient teachers. Deep gratitude to Frank Fish, professor of biology at West Chester University; Mark Blaxter, head of the Tree of Life Programme at the Wellcome Sanger Institute; Judy Müller-Cohn, cofounder of Biomatrica Inc. and COO at BioFluidica; Rolf Müller, cofounder of Biomatrica Inc.; Nikki Smith, ice climbing instructor and photographer; Pamela Silver, biologist and bioengineer at Wyss Institute at Harvard University; John "Griff" Griffith, naturalist at the Humboldt Redwoods State Park; Brook Kennedy, award-winning industrial designer and professor at Virginia Polytechnic Institute; Jonathan Boreyko, who directs Virginia Tech's Nature-Inspired Fluids and Interfaces Lab; Todd Dawson, biologist and professor at University of California, Berkeley; Christian Aalkjær, a professor at Aarhus University and Copenhagen University; Judith "Judy" Racusin, astrophysicist at NASA's Goddard Space Flight Center; Dale DeNardo, attending veterinarian and professor at Arizona State University; Melissa Wilson, a computational biologist at Arizona State University; Judith Kalinyak, a nuclear medicine specialist and endocrinologist; Max Vladymyrov, a machine learning research scientist; Robin Andrews, a fellow science journalist; Nicholas Kotov, Irving Langmuir Professor of Chemical Sciences and Engineering at the University of Michigan; Dr. Fumiya Iida, a researcher at Cambridge's Bio-Inspired Robotics Lab; J. Herbert Waite, a biochemist and professor at UC Santa Barbara who studies marine organisms; Emily Carrington, marine biologist at the University of Washington; Rebecca Johnson, an evolutionary biologist at the California Academy of Sciences; Jeff Brennan, senior vice-president, Global Healthcare Business at Altair; Lamya Karim, bioengineer and assistant professor at the University of Massachusetts Dartmouth; Luke DeGroote, an avian research coordinator from Carnegie Museum of Natural History; Graham Martin, emeritus professor of avian sensory science at the University of Birmingham; Natalia Ocampo-Peñuela, a postdoctoral fellow at ETH Zurich; Lisa Welch, a marketing and sales representative for Arnold Glas; Sean Monkman, senior vice-president of technology development at CarbonCure; Bart Shepherd, director of the Steinhart Aquarium at the California Academy of Sciences; Anthony Brennan, founder of Sharklet Technologies; James (Jimmy) Liao, associate professor of biology at the University of Florida/Whitney Laboratory for Marine Bioscience; Curt Hallberg, a design engineer and CTO of Watreco; and Katia Bertoldi, a professor of applied mechanics at Harvard University.

The Woods Hole Oceanographic Institution in Falmouth, Massachusetts, gave me access to their scientists and labs for a week, even when a bomb cyclone hit the region; roads were blocked and woolen blankets handed out to those of us stuck in the dark, but we made it work. The California Academy of Sciences continues to welcome me with open arms and provide an endless source of inspiration. The Bozeman Ice Festival was a wondrous experience with the most kindhearted, humble people. When I visited Humboldt Redwoods State Park, it was in preparation for a chapter in this book, but I spent another day hiking eighteen miles to experience the Eden of these fog giants. I am deeply grateful these places allowed me to experience their world of science and the natural world.

I am especially fortunate for my thoughtful editor, Linda Pruessen at Greystone Books, copy editor extraordinaire Jess Shulman, and literary agent Mary Krienke at Sterling Lord Literistic. My deepest thanks to my mother, who has provided limitless compassion and encouragement. The kernel of this book's genesis had been roaming the recesses of my mind for a couple of years before it ever made it to an editor's table. Lastly, thank you to my friends and family for providing moral support, for giving feedback on my book, and for understanding when I've had to disappear into the mountains and write for an indeterminate amount of time.

Notes

1. Alexander Graham Bell, "The Telephone. A Lecture," delivered before the Society of Telegraph Engineers, October 31, 1877.
2. Experimental notebook 1:13. See Bell, "The Telephone. A Lecture."
3. Described in Bell's lab notebook on March 10, 1876. See Leonard C. Bruno, "'Mr. Watson, Come Here'—First Release of Bell Papers Goes Online," Library of Congress, April 1999, https://www.loc.gov/loc/lcib/9904/bell.html.
4. Nobel Foundation, "Glowing Proteins—A Guiding Star for Biochemistry," Nobel Prize in Chemistry 2008 press release, October 8, 2008, https://www.nobelprize.org/prizes/chemistry/2008/press-release/.
5. United Nations Office of the High Commissioner for Human Rights in cooperation with the International Bar Association, *Human Rights in the Administration of Justice: A Manual on Human Rights for Judges, Prosecutors and Lawyers*, United Nations, 2003.
6. Anton Leewenhoeck, "An Abstract of a Letter From Mr. Anthony Leewenhoeck at Delft, Dated Sep. 17. 1683," *Philosophical Transactions* 14 (May 20, 1684): 568–74, https://royalsocietypublishing.org/doi/10.1098/rstl.1684.0030.
7. Letter from Antony van Leeuwenhoek to Robert Hooke, November 12, 1680, translated in Clifford Dobell, *Antony van Leeuwenhoek and His Little Animals* (New York: Russell and Russell, 1958), 200.
8. "Microbiology by Numbers," *Nature Reviews Microbiology* 9, no. 628 (August 12, 2011), https://doi.org/10.1038/nrmicro2644.
9. Mark Blaxter, interview with author, August 31, 2020. Unless otherwise attributed, all quotations from Blaxter in this chapter are taken from this interview.
10. Judy Müller-Cohn, interview with author, August 16, 2020. Unless otherwise attributed, all quotations from Müller-Cohn in this chapter are taken from this interview.
11. John H. Crowe and Lois M. Crowe, Method for Preserving Liposomes, U.S. Patent 4,857,319, assigned to the Regents of the University of California, August 15, 1989.
12. Rolf Müller, interview with author, August 16, 2020.
13. Sukee Bennett, "After Conquering Space, Water Bears Could Save the Global Vaccine and Blood Supply," NOVA, PBS, March 6, 2019, https://www.pbs.org/wgbh/nova/article/after-conquering-space-water-bears

-could-save-global-vaccine-and-blood-supply/.

14. Michael Marshall, "Tardigrades: Nature's Great Survivors," *Guardian*, March 20, 2021, https://www.theguardian.com/science/2021/mar/20/tardigrades-natures-great-survivors.
15. United Nations Environment Programme, "About Montreal Protocol," https://www.unep.org/ozonaction/who-we-are/about-montreal-protocol.
16. Quoted in Joe Palca, "Telescope Innovator Shines His Genius on New Fields," *Morning Edition*, NPR, August 23, 2021.
17. Albert Van Helden, "The Invention of the Telescope," *Transactions of the American Philosophical Society* 67, no. 4 (1977): 1–67.
18. Johannes Kepler, *Kepler's Conversation With Galileo's Sidereal Messenger*, trans. Edward Rosen (New York: Johnson Reprint Corp., 1965).
19. Geoff Cottrell, *Telescopes: A Very Short Introduction* (Oxford: Oxford University Press, 2016).
20. "J. Roger Angel, 2016 National Inventors Hall of Fame Inductee," USPTOvideo, YouTube, 2016, https://www.youtube.com/watch?v=x1oaxvlOD9Y.
21. Richard Dawkins, *Climbing Mount Improbable* (New York: W. W. Norton, 1997).
22. Michael F. Land, "Animal Eyes With Mirror Optics," *Scientific American*, December 1978.
23. Beverly Karplus Hartline, "Lobster-Eye X-Ray Telescope Envisioned," *Science* 207, no. 4426 (January 4, 1980).
24. Land, "Animal Eyes With Mirror Optics."
25. Hartline, "Lobster-Eye X-Ray Telescope Envisioned."
26. Judith Racusin, interview with author, November 19, 2020. Unless otherwise attributed, all quotations from Racusin in this chapter are taken from this interview.
27. "Lobster Telescope Has an Eye for X-Rays," University of Leicester press release, April 2006, https://www.le.ac.uk/ebulletin-archive/ ebulletin/news/press-releases/2000-2009/2006/04/nparticle-2n7-yxb-kmd.html.
28. "Mercury Imaging X-Ray Spectrometer (MIXS)," University of Leicester, https://www2.le.ac.uk/departments/physics/research/src/Missions/bepicolombo/mecury-imaging-x-ray-spectrometer-mixs.
29. "Gamma Rays," Science Mission Directorate, NASA Science, 2010, http://science.nasa.gov/ems/12_gammarays.
30. Nan Shepherd, "The Color of Deeside," in *The Deeside Field* 8 (Aberdeen: Aberdeen University Press, 1937), 11.
31. Diane Ackerman, *The Human Age: The World Shaped by Us* (New York: W. W. Norton, 2014).
32. Todd Dawson, interview with author, August 31, 2020.
33. Brook Kennedy, interview with author, August 18, 2020. Unless

otherwise attributed, all quotations from Kennedy in this chapter are taken from this interview.

34. Jonathan Boreyko, interview with author, August 17, 2020. Unless otherwise attributed, all quotations from Boreyko in this chapter are taken from this interview.
35. Deborah Gordon, *Ants at Work: How an Insect Society Is Organized* (New York: W. W. Norton, 2020).
36. Gordon, *Ants at Work*.
37. Eric Bonabeau and Christopher Meyer, "Swarm Intelligence: A Whole New Way to Think About Business," *Harvard Business Review*, May 2001.
38. Friedrich Nietzsche, *Beyond Good and Evil*, trans. R. J. Hollingdale (London: Penguin Publishing Group, 2003); first published in 1886.
39. Rodney A. Brooks and Anita M. Flynn, "Fast, Cheap and Out of Control: A Robot Invasion of the Solar System," *Journal of the British Interplanetary Society* 42 (1989): 478–85.
40. Quoted in Caroline Perry, "A Self-Organizing Thousand-Robot Swarm," Harvard School of Engineering and Applied Sciences press release, August 14, 2014.
41. Ed Yong, "How the Science of Swarms Can Help Us Fight Cancer and Predict the Future," *Wired*, March 19, 2013.
42. Thomas Seeley, S. Kuhnolz, and Robin Hadlock Seeley, "An Early Chapter in Behavioral Physiology and Sociobiology: The Science of Martin Lindauer," *Journal of Comparative Physiology A* 188, no. 6 (August 2002): 439–53.
43. Seeley, "An Early Chapter."
44. Seeley, "An Early Chapter."
45. Thomas D. Seeley, Kevin Passino, and Kirk Visscher, "Group Decision Making in Honey Bee Swarms," *American Scientist* 94, no. 3 (May–June 2006): 220.
46. Clint A. Penick et al., "External Immunity in Ant Societies: Sociality and Colony Size Do Not Predict Investment in Antimicrobials," *Royal Society Open Science* 5, no. 2 (February 2018).
47. Nick Bos et al., "Ants Medicate to Fight Disease," *Evolution* 69, no. 11 (November 2015): 2979–84.
48. Yong, "How the Science of Swarms."
49. "The Giraffe," *Inside Nature's Giants* season 1, episode 4, created by Mark Burnett, developed by 4International, Channel 4, July 2009.
50. Mathew Wedel, "A Monument of Inefficiency: The Presumed Course of the Recurrent Laryngeal Nerve in Sauropod Dinosaurs," *Acta Palaeontologica Polonica* 57, no. 2 (June 2012).
51. August Krogh, "The Progress of Physiology," *American Journal of Physiology* 90, no. 2 (October 1, 1929), https://doi.org/10.1152/ajplegacy.1929.90.2.243.
52. Christian Aalkjær, interview with author, August 26, 2020. Unless otherwise attributed, all quotations from Aalkjær in this chapter are taken from this interview.

53. "Hertha's Story," *Vein Magazine*, April 1, 2015, https://www.veindirectory.org/magazine/article/industry-spotlight/herthas-story.
54. Joseph F. Kubis et al., "Apollo 15, Time and Motion Study," National Aeronautics and Space Administration Manned Spacecraft Center, January 1972.
55. Joseph F. Kubis et al., "Apollo 16, Time and Motion Study," National Aeronautics and Space Administration Manned Spacecraft Center, July 1972.
56. Diana Young and Dava Newman, "Augmenting Exploration: Aerospace, Earth and Self," in *Wearable Monitoring Systems*, eds. Annalisa Bonfiglio and Danilo De Rossi (Boston: Springer, 2011); chapter available online at https://doi.org/10.1007/978-1-4419-7384-9_11.
57. J. Herbert Waite, interview with author, December 9, 2020. Unless otherwise attributed, all quotations from Waite in this chapter are taken from this interview.
58. Thomas Lambert, *Bone Products and Manures: A Treatise on the Manufacture of Fat, Glue, Animal Charcoal, Size, Gelatin, and Manures* (London: Scott, Greenwood, 1925).
59. Valeria J. Brown, "Better Bonding With Beans," *Environmental Health Perspectives* 113, no. 8 (August 2005): A538–41.
60. Brown, "Better Bonding With Beans."
61. Keith Hautala, "We Make Innovations That Stick: Sustainable Adhesives," Oregon State University, December 18, 2019, https://cbee.oregonstate.edu/node/825.
62. Nick Houtman, "Oregon State University Researcher Receives National Award for Soy-Based Adhesive," Oregon State University, November 8, 2017, https://today.oregonstate.edu/news/oregon-state-university-researcher-receives-national-award-soy-based-adhesive.
63. Emily Carrington, interview with author, December 3, 2020. Unless otherwise attributed, all quotations from Carrington in this chapter are taken from this interview.
64. Jill Needham and Kathleen Alcalá, "Mussel Strength: How Mussels Serve Our Ecosystem," *Salish Magazine*, n.d., https://salishmagazine.org/mussel-strength/.
65. Bart Shepherd, interview with author, November 8, 2018.
66. Quoted at Breakthrough Energy, https://www.breakthroughenergy.org.
67. Bill Gates, *How to Avoid a Climate Disaster* (New York: Alfred A. Knopf, 2021).
68. Robert J. Gordon, *The Rise and Fall of American Growth* (New Jersey: Princeton University Press, 2017).
69. "Edison and the Electric Car," Thomas A. Edison Papers, Rutgers School of Arts and Sciences, October 28, 2016, http://edison.rutgers.edu/elecar.htm.

70. "Lithium-Ion Batteries," scientific background on the Nobel Prize in Chemistry 2019, Royal Swedish Academy of Sciences, October 9, 2019, https://www.nobelprize.org/uploads/2019/10/advanced-chemistryprize2019.pdf.

71. Nian Liu et al., "A Pomegranate-Inspired Nanoscale Design for Large-Volume-Change Lithium Battery Anodes," *Nature Nanotechnology* 9, no. 3 (March 2014): 187–92, https://www.nature.com/articles/nnano.2014.6.

72. Angela Belcher, "Using Nature to Grow Batteries," TEDx Talk, April 2011, https://www.ted.com/talks/angela_belcher_using_nature_to_grow_batteries.

73. Source for "androids" is Khan Academy, "Are Viruses Dead or Alive?," https://www.khanacademy.org/test-prep/mcat/cells/viruses/a/are-viruses-dead-or-alive; source for "microscopic zombies" is Daniel Oberhaus, "The Next Generation of Batteries Could Be Built by Viruses," *Wired*, February 26, 2020, https://www.wired.com/story/the-next-generation-of-batteries-could-be-built-by-viruses.

74. "Nicholas A. Kotov," Chemical Engineering, University of Michigan, https://che.engin.umich.edu/people/kotov-nicholas/.

75. Nicholas Kotov, interview with author, March 11, 2021. Unless otherwise attributed, all quotations from Kotov in this chapter are taken from this interview.

76. Fumiya Iida, interview with author, March 16, 2021. Unless otherwise attributed, all quotations from Iida in this chapter are taken from this interview.

77. "Electric Vehicle Sales to Fall 18% in 2020 but Long-term Prospects Remain Undimmed," BloombergNEF, May 19, 2020, https://about.bnef.com/blog/electric-vehicle-sales-to-fall-18-in-2020-but-long-term-prospects-remain-undimmed/.

78. Quoted in Mary Fagan, "Sheikh Yamani Predicts Price Crash as Age of Oil Ends," *Telegraph*, June 25, 2000, https://www.telegraph.co.uk/news/uknews/1344832/Sheikh-Yamani-predicts-price-crash-as-age-of-oil-ends.html.

79. Jeff Brennan, interview with author, October 12, 2020. Unless otherwise attributed, all quotations from Brennan in this chapter are taken from this interview.

80. Leonardo da Vinci, *Leonardo da Vinci's Notebooks: Arranged and Rendered Into English With Introductions*, trans. Edward McCurdy (London: Duckworth and Company, 1908).

81. This example is outlined by Michael Fowler, physics professor at the University of Virginia, at https://galileoandeinstein.phys.virginia.edu/lectures/scaling.html.

82. Galileo Galilei, *Dialogues Concerning Two New Sciences*, trans. Alfonso de Salvio and Henry Crew (New York: Macmillan Company, 1914).

83. "Origins and Construction of the Eiffel Tower," Eiffel Tower tourism website, https://www.toureiffel.paris/en/the-monument/history; and Aatish Bhatia, "What Your Bones Have in Common With the Eiffel Tower," *Wired*, March 9, 2015, https://www.wired.com/2015/03/empzeal-eiffel-tower/.

84. Cleve Wootson Jr., "'I Guess You Are Here for the Opium': Investigator Stumbles Across $500 Million in Poppy Plants," *Washington Post*, May 25, 2017, https://www.washingtonpost.com/news/to-your-health/wp/2017/05/25/i-guess-you-are-here-for-the-opium-investigator-stumbles-across-500-million-in-poppy-plants/.

85. S. Y. Tan and Y. Tatsumura, "Alexander Fleming (1881–1955): Discoverer of Penicillin," *Singapore Medical Journal* 56, no. 7 (2015): 366–67, https://doi.org/10.11622/smedj.2015105.

86. George Washington Tyron, Henry Augustus Pilsbry, and B. Sharp, *Manual of Conchology, Structural and Systematic*, vol. 6, *Conidae, Pleurotomidae* (Philadelphia: Academy of Natural Sciences, 1884).

87. Ross Piper et al., "Nature Is a Rich Source of Medicine—If We Can Protect It," *The Conversation*, December 14, 2021, https://theconversation.com/nature-is-a-rich-source-of-medicine-if-we-can-protect-it-107471.

88. "The Gila Monster: A Rare Reptile Described as the Most Venomous Thing on Earth," *Newberry Herald and News*, October 23, 1890, https://www.newspapers.com/image/174576495/; and "Ugly as Sin Itself," *San Francisco Chronicle*, June 18, 1893, https://www.newspapers.com/image/27601356/.

89. Earl F. Nation, "George E. Goodfellow, M.D. (1855–1910): Gunfighter's Surgeon and Urologist," *Urology* 2, no. 1 (1973): 85–92, https://doi.org/10.1016/0090-4295(73)90226-4.

90. George Goodfellow, "The Gila Monster Again," *Scientific American* 96, no. 13 (1907), 271, http://www.jstor.org/stable/26005510.

91. Dale DeNardo, interview with author, June 2, 2020. Unless otherwise attributed, all quotations from DeNardo in this chapter are taken from this interview.

92. Ralph Waldo Emerson, *The Prose Works of Ralph Waldo Emerson: Representative Men. English Traits. Conduct of Life* (Boston: J. R. Osgood and Company, 1872).

93. "The Worldwide Rise of Chronic Noncommunicable Diseases: A Slow-Motion Catastrophe," World Health Organization, https://www.who.int/director-general/speeches/detail/the-worldwide-rise-of-chronic-noncommunicable-diseases-a-slow-motion-catastrophe.

94. Judith Kalinyak, interview with author, June 20, 2021.

95. Denise Gellene, "Lizard Is Source of Newest Diabetes

Drug," *Los Angeles Times*, April 30, 2005, https://www.latimes.com/archives/la-xpm-2005-apr-30-fi-gila30-story.html.

96. Andrew Pollack, "Lizard-Linked Therapy Has Roots in the Bronx," *New York Times*, September 21, 2002, https://www.nytimes.com/2002/09/21/business/lizard-linked-therapy-has-roots-in-the-bronx.html.

97. Charles Darwin, *On the Origin of Species* (London: John Murray, 1859).

98. Frank Fish, interview with author, August 26, 2020. Unless otherwise attributed, all quotations from Fish in this chapter are taken from this interview.

99. John Gierach, *Trout Bum* (Berkeley: West Margin Press, 2013).

100. James "Jimmy" Liao, interview with author, August 19, 2020. Unless otherwise attributed, all quotations from Liao in this chapter are taken from this interview.

101. "Michael Bernitsas," Naval Architecture and Marine Engineering faculty bio, University of Michigan, n.d., https://name.engin.umich.edu/people/bernitsas-michael/.

102. Anna Gruener, "The Effect of Cataracts and Cataract Surgery on Claude Monet," *British Journal of General Practice* 65, no. 634 (2015): 254–55, https://bjgp.org/content/65/634/254.

103. Scott R. Loss et al., "Bird–Building Collisions in the United States: Estimates of Annual Mortality and Species Vulnerability," *The Condor* 116, no. 1 (February 1, 2014): 8–23, https://doi.org/10.1650/CONDOR-13-090.1.

104. Luke DeGroote, interview with author, September 1, 2020. Unless otherwise attributed, all quotations from DeGroote in this chapter are taken from this interview.

105. "Birds Eat 400 to 500 Million Tons of Insects Annually," *ScienceDaily*, July 9, 2018, https://www.sciencedaily.com/releases/2018/07/180709100850.htm.

106. Lisa Welch, interview with author, August 28, 2020. Unless otherwise attributed, all quotations from Welch in this chapter are taken from this interview.

107. Graham Martin, interview with author, August 27, 2020. Unless otherwise attributed, all quotations from Martin in this chapter are taken from this interview.

108. Heath Waldorf, "What the New Bird Friendly Local Law 15 Means to You," *NYREJ*, June 23, 2020, https://nyrej.com/what-the-new-bird-friendly-local-law-15-means-to-you-by-heath-waldorf.

109. Jennifer Chu, "Super-Strong Surgical Tape Detaches on Demand," *MIT News*, June 22, 2020, https://news.mit.edu/2020/surgical-tape-wounds-0622.

110. Qi Guo et al., "Compact Single-Shot Metalens Depth Sensors Inspired by Eyes of Jumping Spiders," *Proceedings of the National Academy of Sciences* 116, no. 46 (2019): 22959–65, https://doi.org/10.1073/pnas.1912154116.

111. Osamu Shimomura, Sachi Shimomura, and John H.

Brinegar, *Luminous Pursuit: Jellyfish, GFP, and the Unforeseen Path to the Nobel Prize* (Singapore: World Scientific, 2017).

112. Nobel Foundation, "Glowing Proteins."

113. Nobel Foundation, "Glowing Proteins."

114. Martin Chalfie, "GFP: Lighting Up Life," Nobel lecture, December 8, 2008.

115. Osamu Shimomura, "Discovery of Green Fluorescent Protein (GFP)," Nobel lecture, December 8, 2008.

116. Shimomura, "Discovery of Green Fluorescent Protein."

117. Benjamin Franklin, April 23, 1752, in *The Works of Dr. Benjamin Franklin: Philosophical. Essays and Correspondence* (London: John Sharpe, 1809), 73.

118. Pliny the Elder, *The Natural History of Pliny*, trans. Henry Thomas Riley and John Bostock, vol. 6 (London: George Bell and Sons, 1989).

119. Edmund Newton Harvey, *A History of Luminescence: From the Earliest Times Until 1900* (Philadelphia: American Philosophical Society, 1957).

120. Charles Leonard Hogue, *Latin American Insects and Entomology* (Berkeley: University of California Press, 1993).

121. Harvey, *A History of Luminescence.*

122. Chalfie, "GFP: Lighting Up Life."

123. "2008 Nobel Prize in Chemistry Shared by UC San Diego Researcher Roger Tsien," University of California San Diego Health press release, October 8, 2008.

124. Roger Tsien, "Unlocking Cell Secrets With Light Beams and Molecular Spies," acceptance speech, Heineken Prize for Biochemistry and Biophysics, 2002.

125. Quoted in Liz Karagianis, "The Brilliance of Basic Research," *MIT Spectrum*, Spring 2014.

126. Shimomura et al., *Luminous Pursuit.*

Bibliography

1: A Cold Case

Agronis, Amy. "Center Works to Bring Life to That Which Is 'Dead.'" UC Davis, November 15, 2004. https://www.ucdavis.edu/news/center-works-bring-life-which-'dead'.

Bennett, Sukee. "After Conquering Space, Water Bears Could Save the Global Vaccine and Blood Supply." *NOVA*, PBS, March 6, 2019. https://www.pbs.org/wgbh/nova/article/after-conquering-space-water-bears-could-save-global-vaccine-and-blood-supply/.

Blaxter, Mark. Interview with author, August 31, 2020.

Blow, N. "Biobanking: Freezer Burn." *Nature Methods* 6, no. 2 (2009): 173–178.

Boothby, T. C., and G. J. Pielak. "Intrinsically Disordered Proteins and Desiccation Tolerance: Elucidating Functional and Mechanistic Underpinnings of Anhydrobiosis." *Bioessays* 39, no. 11 (November 2017). https://onlinelibrary.wiley.com/doi/abs/10.1002/bies.201700119.

Boothby, T. C., H. Tapia, A. H. Brozena, S. Piszkiewicz, A. E. Smith, I. Giovannini, L. Rebecchi, G. J. Pielak, D. Koshland, and B. Goldstein. "Tardigrades Use Intrinsically Disordered Proteins to Survive Desiccation." *Molecular Cell* 65, no. 6 (2017): 975–84.

Boothby, Thomas. "Ted-Ed: Meet the Tardigrade, the Toughest Animal on Earth." TEDx Talk, 2017. https://www.ted.com/talks/thomas_boothby_meet_the_tardigrade_the_toughest_animal_on_earth.

Crowe, J. H., J. F. Carpenter, and L. M. Crowe. "The Role of Vitrification in Anhydrobiosis." *Annu Rev Physiol* 60, no. 1 (1998): 73–103.

Crowe, John H., and Lois M. Crowe. Method for Preserving Liposomes. U.S. Patent 4,857,319, August 15, 1989.Assignee: The Regents of the University of California.

Crowe, L. M. "Lessons From Nature: The Role of Sugars in Anhydrobiosis." *Comp Biochem Physiol A Mol Integr Physiol* 131, no. 3 (2002): 505–13.

Czernekova, M., and K. I. Jonsson. "Experimentally Induced Repeated Anhydrobiosis in the Eutardigrade Richtersius Coronifer." *PLoS One* 11, no. 11 (2016). https://pubmed.ncbi.nlm.nih.gov/27828978/.

Howard, L. "Research Inspired by 'Water Bears' Leads to Innovations in Medicine, Food Preservation and Blood Storage." UC Davis, August 5, 2019. https://research.ucdavis.edu/research-inspired-by-water-bears-leads-to-innovations-in-medicine-food-preservation-and-blood-storage/.

Jonsson, K. I. "Radiation Tolerance in Tardigrades: Current Knowledge and Potential Applications in Medicine." *Cancers* 11, no. 9 (2019): 1333.

Jonsson, K. I., E. Rabbow, R. O. Schill, M. Harms-Ringdahl, and P. Rettberg. "Tardigrades Survive Exposure to Space in Low Earth Orbit." *Current Biology* 18, no. 17 (2008): R729–31.

Leeuwenhoek, Antony van, to Robert Hooke, letter, November 12, 1680, translated in *Antony van Leeuwenhoek and His Little Animals*, by C. Dobell. New York: Russell and Russell, 1958.

Leewenhoeck, Anton. "An Abstract of a Letter From Mr. Anthony Leewenhoeck at Delft, Dated Sep. 17. 1683." *Philosophical Transactions* 14 (May 20, 1684): 568–74. https://royalsocietypublishing.org/doi/10.1098/rstl.1684.0030.

Leslie, S. B., E. Israeli, B. Lighthart, J. H. Crowe, and L. M. Crowe. "Trehalose and Sucrose Protect Both Membranes and Proteins in Intact Bacteria During Drying." *Appl Environ Microbiol* 61, no. 10 (1995): 3592–97.

Lin, Q., Q. Zhao, and B. Lev. "Cold Chain Transportation Decision in the Vaccine Supply Chain." *European Journal of Operational Research* 283, no. 1 (2020): 182–95.

Marshall, Michael. "Tardigrades: Nature's Great Survivors." *Guardian*, March 20, 2021. https://www.theguardian.com/science/2021/mar/20tardigrades-natures-great-survivors.

"Microbiology by Numbers." *Nat Rev Microbiol* 9, no. 628 (2011). https://doi.org/10.1038/nrmicro2644.

Moeti, M., R. Nandy, S. Berkley, S. Davis, and O. Levine. "No Product, No Program: The Critical Role of Supply Chains in Closing the Immunization Gap." *Vaccine* 35, no. 17 (2017): 2101–2.

Mogle, M. J., S. A. Kimball, W. R. Miller, and R. D. McKown. "Evidence of Avian-Mediated Long Distance Dispersal in American Tardigrades." *PeerJ* 6 (July 4, 2018): e5035.

Müller, Rolf. Interview with author, August 16, 2020.

Müller-Cohn, Judy. Interview with author, August 16, 2020.

Rogers, B., K. Dennison, N. Adepoju, S. Dowd, and K. Uedoi. "Vaccine Cold Chain: Part 1. Proper Handling and Storage of Vaccine." AAOHN *Journal* 58, no. 9 (2010): 337–46.

Tsujimoto, M., S. Imura, and H. Kanda. "Recovery and Reproduction of an Antarctic Tardigrade Retrieved From a Moss Sample Frozen for Over 30 Years." *Cryobiology* 72, no. 1 (2016): 78–81.

United Nations Environment Programme. "About Montreal Protocol." N.d. https://www.unep.org/ozonaction/who-we-are/about-montreal-protocol.

Westover, C., D. Najjar, C. Meydan, K. Grigorev, M. Veling, R. Chang, S. Iosim, et al. "Engineering Radioprotective Human Cells Using the Tardigrade Damage Suppressor Protein, DSUP." *BioRxiv* (2020), https://doi.org/10.1101/2020.11.10.373571.

Woolhouse, M., F. Scott, Z. Hudson, R. Howey, and M. Chase-Topping. "Human Viruses: Discovery and Emergence." *Philosophical Transactions of the Royal Society B: Biological Sciences* 367, no. 1604 (2012): 2864–71.

World Health Organization and UNICEF. "Progress and Challenges With Achieving Universal Immunization Coverage." July 2020. https://www.who.int/immunization/monitoring_surveillance/who-immuniz.pdf?ua=1.

Yin, S. "Searching Tardigrades for Lifesaving Secrets." *New York Times*, February 15, 2019. https://www.nytimes.com/2019/02/15/health/tardigrades-suspended-animation.html.

2: Fishing for Stars

Ackerman, Diane. *The Human Age: The World Shaped by Us*. New York: W. W. Norton, 2014.

Angel, J. R. "Lobster Eyes as X-Ray Telescopes." *SPIE Proceedings*, 1979. https://doi.org/10.1117/12.957437.

Billings, Lee. "Catching the Stars." *Aeon*, 2013. https://aeon.co/essays/when-this-man-talks-about-energy-the-world-needs-to-listen.

Cottrell, Geoff. *Telescopes: A Very Short Introduction*. Oxford: Oxford University Press, 2016.

Dawkins, Richard. *Climbing Mount Improbable*. New York: W. W. Norton, 1997.

Fisher, Arthur. "Spinning Scopes." *Popular Science*, October 1987.

Gotz, D., C. Adami, S. Basa, V. Beckmann, V. Burwitz, R. Chipaux, B. Cordier, et al. "The Microchannel X-Ray Telescope on Board the SVOM Satellite." *ArXiv* preprint, 2015. https://arxiv.org/abs/1507.00204.

Greanya, V. *Bioinspired Photonics: Optical Structures and Systems Inspired by Nature*. Boca Raton, FL: CRC Press, 2015.

Hartline, Beverly Karplus. "Lobster-Eye X-Ray Telescope Envisioned." *Science* 207, no. 4426 (January 4, 1980).

Helden, Albert Van. "The Invention of the Telescope." *Transactions of the American Philosophical Society* 67, no. 4 (1977): 1–67.

Hill, John M., James Roger P. Angel, Randall D. Lutz, Blain H. Olbert, and Peter A. Strittmatter. "Casting the First 8.4-m Borosilicate Honeycomb Mirror for the Large Binocular Telescope." *SPIE Proceedings*, 1998. https://doi.org/10.1117/12.319295.

Hudec, R., L. Pina, V. Simon, L. Sveda, A. Inneman, V. Semencova, and M. Skulinova. "Lobster: New Space X-Ray Telescopes." *Nuclear Physics B-Proceedings Supplements* 166 (2007): 229–33.

Keesey, L. "Measuring Transient X-Rays With Lobster Eyes." NASA, May 2012. https://www.nasa.gov/topics/technology/features/lobster-eyes.html.

Kepler, Johannes. *Kepler's Conversation With Galileo's Sidereal Messenger*. Trans. Edward Rosen. New York: Johnson Reprint Corp., 1965.

Land, Michael F. "Animal Eyes With Mirror Optics." *Scientific American*, December 1978.

———. *Eyes to See: The Astonishing Variety of Vision in Nature*. Oxford: Oxford University Press, 2018.

Land, Michael F., and Dan-Eric Nilsson. *Animal Eyes*. Oxford: Oxford University Press, 2012.

Lankford, J. *History of Astronomy: An Encyclopedia*. Oxfordshire, U.K.: Routledge, 2013.

Palca, Joe. "Telescope Innovator Shines His Genius on New Fields." *Morning Edition*, NPR, August 23, 2012. https://www.npr.org/2012/08/23/159554100/telescope-innovator-shines-his-genius-on-new-fields.

Photonis. "Space Qualified Imaging" (product brochure). 2017. https://www.photonis.com/system/files/2019-03/Micro-Pore-Optic-brochure.pdf.

Racusin, Judith. Interview with author, November 19, 2020.

Science Mission Directorate. "Gamma Rays." NASA Science, 2010.

https://science.nasa.gov/ems/12_gammarays.

Shepherd, Nan. "The Color of Deeside." In *The Deeside Field* 8:11. Aberdeen: Aberdeen University Press, 1937.

University of Leicester. "Lobster-Inspired £3.8m Super Lightweight Mirror Chosen for Chinese–French Space Mission." *PhysOrg*, October 26, 2015. https://phys.org/news/2015-10-lobster-inspired-38m-super-lightweight-mirror.html.

———. "Lobster Telescope Has an Eye for X-Rays." Press release, April 2006. https://www.le.ac.uk/ebulletin-archive/ebulletin/news/press-releases/2000-2009/2006/04/nparticle-2n7-yxb-kmd.html.

———. "Mercury Imaging X-Ray Spectrometer (MIXS)." N.d. https://www2.le.ac.uk/departments/physics/research/src/Missions/bepicolombo/mecury-imaging-x-ray-spectrometer-mixs.

Urban, M., O. Nentvich, V. Stehlikova, T. Baca, V. Daniel, and R. Hudec. "Vzlusat-1: Nanosatellite With Miniature Lobster Eye X-Ray Telescope and Qualification of the Radiation Shielding Composite for Space Application." *Acta Astronautica* 140 (2017): 96–104.

USPTOvideo. "J. Roger Angel, 2016 National Inventors Hall of Fame Inductee." YouTube video, 2016. https://www.youtube.com/watch?v=x10axvlOD9Y.

Vukusic, P., and J. R. Sambles. "Photonic Structures in Biology." *Nature* 424, no. 6950 (2003): 852–55.

3: Drinking From a Cloud

Boreyko, Jonathan. Interview with author, August 17, 2020.

Boreyko, Jonathan B., and Chuan-Hua Chen. "Restoring Superhydrophobicity of Lotus Leaves With Vibration-Induced Dewetting." *Physical Review Letters* 103, no. 17 (2009). https://doi.org/10.1103/physrevlett.103.174502.

———. "Self-Propelled Dropwise Condensate on Superhydrophobic Surfaces." *Physical Review Letters* 103, no. 18 (2009). https://doi.org/10.1103/physrevlett.103.184501.

Burgess, Stephen, and Todd Dawson. "The Contribution of Fog to the Water Relations of *Sequoia sempervirens* (D. Don): Foliar Uptake and Prevention of Dehydration." *Plant, Cell & Environment* (2004): 1023–34.

California's Redwood State Parks. Informational brochure, 2017. https://www.parks.ca.gov/pages/24723/files/CARedwoodsSPFinalWebLayout2017.pdf.

Dawson, Todd. "Fog in the California Redwood Forest: Ecosystem Inputs and Use by Plants." *Oecologia* 117, no. 4 (1998): 476–85.

———. Interview with author, August 31, 2020.

Dokter, A. M., A. Farnsworth, D. Fink, V. Ruiz-Gutierrez, W. M. Hochachka, F. A. La Sorte, O. J. Robinson, K. V. Rosenberg, and S. Kelling. "Seasonal Abundance and Survival of North America's Migratory Avifauna Determined by Weather Radar." *Nature Ecology & Evolution* 2, no. 10 (2018): 1603–9.

Gould, P. "Smart, Clean Surfaces." *Materials Today* 6, no. 11 (2003): 44–48.

Kennedy, Brook. Interview with author, August 18, 2020.

Kennedy, Brook, Jonathan Boreyko, and Weiwei Shi. "Fog Harp: University Invention to Real-World Impact." *Zygote Quarterly* 1, no. 27 (2020). https://issuu.com/eggermont/docs/zq_issue_27.

Kennedy, Brook, Jonathan Boreyko, Weiwei Shi, M. Anderson, J. Tulkoff, and T. Van der Sloot. "Designing a Fog-Harvesting Harp." Industrial Designers Society of America, n.d. https://www.idsa.org/sites/default/files/FINAL_Paper_Designing%20a%20Fog-Harvesting%20Harp.pdf.

Mekonnen, Mesfin, and Arjen Hoekstra. "Four Billion People Facing Severe Water Scarcity." *Science Advances* 2, no. 2 (February 12, 2016).

Nuwer, R. "In Towering Redwoods, an Abundance of Tiny, Unseen Life." *New York Times,* April 19, 2016. https://www.nytimes. com/2016/04/19/science/in-towering-redwoods-an-abundance-of-tiny-unseen-life.html.

Preston, R. *The Wild Trees: A Story of Passion and Daring.* New York: Random House, 2008.

Roediger, E. "Out of the Lab, Into the Field." *Virginia Tech Engineer,* Fall 2018. https://eng.vt.edu/magazine/stories/fall-2018/fog-harp.html.

Shi, W., M. J. Anderson, J. B. Tulkoff, B. S. Kennedy, and J. B. Boreyko. "Fog Harvesting With Harps." *ACS Applied Materials & Interfaces* 10, no. 14 (2018): 11979–86.

Shi, W., T. W. van der Sloot, B. J. Hart, B. S. Kennedy, and J. B. Boreyko. "Harps Enable Water Harvesting Under Light Fog Conditions." *Advanced Sustainable Systems* 4, no. 6 (2020). https://onlinelibrary.wiley.com/doi/full/10.1002/adsu.202000040.

United Nations. *UN World Water Development Report*. 2014. https://www.unwater.org/publications/world-water-development-report-2014-water-energy/.

Xu, Q., W. Zhang, C. Dong, T. S. Sreeprasad, and Z. Xia. "Biomimetic Self-Cleaning Surfaces: Synthesis, Mechanism and Applications." *Journal of the Royal Society Interface* 13, no. 122 (September 1, 2016). https://doi.org/10.1098/rsif.2016.0300.

Xu, Q., Y. Wan, T. S. Hu, T. X. Liu, D. Tao, P. H. Niewiarowski, Y. Tian, et al. "Robust Self-Cleaning and Micromanipulation Capabilities of Gecko Spatulae and Their Bio-Mimics." *Nature Communications* 6, no. 1 (2015): 1–9.

Zhai, Lei, Michael C. Berg, Fevzi Ç. Cebeci, Yushan Kim, John M. Milwid, Michael F. Rubner, and Robert E. Cohen. "Patterned Superhydrophobic Surfaces: Toward a Synthetic Mimic of the Namib Desert Beetle." *Nano Letters* 6, no. 6 (2006): 1213–17. https://doi.org/10.1021/nl060644q.

4: Who's in Charge?

Angle, C., and R. Brooks. "Small Planetary Rovers." Proceedings of the International Conference on Intelligent Robots and Systems, Tsuchiura, Japan, 1990.

Bares, John E., and David S. Wettergreen. "Dante II: Technical Description, Results, and Lessons Learned." *International Journal of Robotics Research* 18, no. 7 (July 1999): 621–49. https://www.ri.cmu.edu/pub_files/pub2/bares_john_1999_1/bares_john_1999_1.pdf.

Bonabeau, Eric, and Christopher Meyer. "Swarm Intelligence: A Whole New Way to Think About Business." *Harvard Business Review*, May 2001. https://hbr.org/2001/05/swarm-intelligence-a-whole-new-way-to-think-about-business.

Bonabeau, Eric, and Guy Theraulaz. "Swarm Smarts." *Scientific American* 282, no. 3 (2000): 72–79.

Bos, Nick, Liselotte Sundström, Siiri Fuchs, and Dalial Freitak. "Ants Medicate to Fight Disease." *Evolution* 69, no. 11 (2015): 2979–84.

Brooks, Rodney A., and Anita M. Flynn. "Fast, Cheap and Out of Control: A Robot Invasion of the Solar System." *Journal of the British Interplanetary Society* 42 (1989): 478–85.

Centibots Project, 2002. http://www.ai.sri.com/centibots/.

Encycle. "National Retail Chain Scores Seven-Figure Energy Savings." Case study, 2019. https://www.encycle.com/wp-content/uploads/2019/04/Large-Box-Retail-Case-Study.pdf.

Gordon, D. M. *Ants at Work: How an Insect Society Is Organized*. New York: W. W. Norton, 2020.

———. "The Rewards of Restraint in the Collective Regulation of Foraging by Harvester Ant Colonies." *Nature* 498, no. 7452 (2013): 91–93.

Mitchell, Melanie. *Complexity: A Guided Tour*. Oxford: Oxford University Press, 2009.

Nietzsche, Friedrich. *Beyond Good and Evil*. Trans. R. J. Hollingdale. London: Penguin Publishing Group, 2003; originally published in 1886.

Ocko, S. A., H. King, D. Andreen, P. Bardunias, J. S. Turner, R. Soar, and L. Mahadevan. "Solar-Powered Ventilation of African Termite Mounds." *Journal of Experimental Biology* 220, no. 18 (2017): 3260–69.

Pagels, Heinz. *The Dreams of Reason*, 12. New York: Simon & Schuster, 1988.

Pathak, A., S. Kett, and M. Marvasi. "Resisting Antimicrobial Resistance: Lessons From Fungus Farming Ants." *Trends in Ecology & Evolution* 34, no. 11 (2019): 974–76.

Penick, Clint A., Omar Halawani, Bria Pearson, Stephanie Mathews, Margarita M. López-Uribe, Robert R. Dunn, and Adrian A. Smith. "External Immunity in Ant Societies: Sociality and Colony Size Do Not Predict Investment in Antimicrobials." *Royal Society Open Science* 5, no. 2 (2018).

Perry, Caroline. "A Self-Organizing Thousand-Robot Swarm." Harvard School of Engineering and Applied Sciences, August 14, 2014. https://www.seas.harvard.edu/news/2014/08/self-organizing-thousand-robot-swarm.

Seeley, T., S. Kuhnholz, and R. Seeley. "An Early Chapter in Behavioral Physiology and Sociobiology: The Science of Martin Lindauer." *Journal of Comparative Physiology A* 188, no. 6 (2002): 439–53.

Seeley, T. D., K. M. Passino, and P. K. Visscher. "Group Decision Making in Honey Bee Swarms: When 10,000 Bees Go House Hunting, How Do They Cooperatively Choose Their New Nesting Site?" *American Scientist* 94, no. 3 (2006): 220–29.

Yong, Ed. "How the Science of Swarms Can Help Us Fight Cancer and Predict the Future." *Wired*, March 19, 2013. https://www.wired.com/2013/03/powers-of-swarms/.

Zimmer, C. "These Ants Use Germ-Killers, and They're Better Than Ours." *New York Times*, September 26, 2019. https://www. nytimes.com/2019/09/26/science/ants-fungus-antibiotic-resistance.html.

5: A Leggy Turn of Events

Aalkjær, Christian. Interview with author, August 26, 2020.

Agaba, M., E. Ishengoma, W. C. Miller, B. C. McGrath, C. N. Hudson, O. C. B. Reina, A. Ratan, et al. "Giraffe Genome Sequence Reveals Clues to Its Unique Morphology and Physiology." *Nature Communications* 7, no. 1 (2016): 1–8.

Angier, N. "Our Understanding of Giraffes Does Not Measure Up." *New York Times*, October 7, 2014. https://www.nytimes.com/2014/10/07/science/our-understanding-of-giraffes-does-not-measure-up.html.

Burnett, Mark. "The Giraffe." *Inside Nature's Giants* season 1, episode 4. Developed by 4International. Channel 4, July 2009.

Chu, J. "Shrink-Wrapping Spacesuits." *MIT News*, September 18, 2014. https://news.mit.edu/2014/second-skin-spacesuits-0918.

Hargens, A. R., R. W. Millard, K. Pettersson, and K. Johansen. "Gravitational Haemodynamics and Oedema Prevention in the Giraffe." *Nature* 329, no. 6134 (1987): 59–60.

"Hertha Peterson Shaw" (obituary). *Coronado Eagle & Journal*, August 19, 2011. http://www.coronadonewsca.com/obituaries/hertha-peterson-shaw/article_329693a4-ca98-11e0-be3e-001cc4c03286.html

"Hertha Shaw: The Inspiration Behind CircAid." CircAid Medical Products, 2011. http://elisesdesigns.com/links/eflash/2011_HerthaStory.html.

"Hertha's Story." *Vein Magazine*, April 1, 2015. https://www.veindirectory.org/magazine/article/industry-spotlight/herthas-story.

Kluger, J., and J. Lovell. *Lost Moon: The Perilous Voyage of Apollo 13*. Boston: Houghton Mifflin, 1994.

Krogh, August. "The Progress of Physiology." *American Journal of Physiology* 90, no. 2 (1929).

Kubis, Joseph F., John T. Elrod, Rudolph Rusnak, and John E. Barnes. "Apollo 15, Time and Motion Study." NASA Manned Spacecraft Center, Houston, Texas, January 1972.

Kubis, Joseph F., John T. Elrod, Rudolph Rusnak, John E. Barnes, and S. Saxon. "Apollo 16, Time and Motion Study (Final Mission Report)."

NASA Manned Spacecraft Center, Houston, Texas, July 1972.

Lydgate, A. "How the Giraffe Got Its Neck." *New Yorker*, May 17, 2016. https://www.newyorker.com/tech/annals-of-technology/how-the-giraffe-got-its-neck.

Micheva, Kristina D., Brad Busse, Nicholas C. Weiler, Nancy O'Rourke, and Stephen J. Smith. "Single-Synapse Analysis of a Diverse Synapse Population: Proteomic Imaging Methods and Markers." *Neuron* 68, no. 4 (November 18, 2010): 639–53.

Newman, D. "Building the Future Spacesuit." *Ask Magazine* 45 (2012): 37–40.

Newman, D. J., M. Canina, and G. L. Trotti. "Revolutionary Design for Astronaut Exploration—Beyond the Bio-Suit System." *American Institute of Physics Conference Proceedings* 880 (2007): 975–86.

Petersen, K. K., A. Hørlyck, K. H. Østergaard, J. Andresen, T. Broegger, N. Skovgaard, N. Telinius, et al. "Protection Against High Intravascular Pressure in Giraffe Legs." *American Journal of Physiology-Regulatory, Integrative and Comparative Physiology* 305, no. 9 (November 1, 2013): R1021–30.

Roth, A. "Dwarf Giraffe Discovery Surprises Scientists." *New York Times*, January 6, 2011. https://www.nytimes.com/2021/01/06/science/dwarf-giraffes.html.

Smerup, M., M. Damkjær, E. Brøndum, U. T. Baandrup, S. B. Kristiansen, H. Nygaard, J. Funder, et al. "The Thick Left Ventricular Wall of the Giraffe Heart Normalises Wall Tension, but Limits Stroke Volume and Cardiac Output." *Journal of Experimental Biology* 219, no. 3 (2016): 457–63.

Warren, James V. "The Physiology of the Giraffe." *Scientific American* 231, no. 5 (1974): 96–105.

Wedel, Mathew. "A Monument of Inefficiency: The Presumed Course of the Recurrent Laryngeal Nerve in Sauropod Dinosaurs." *Acta Palaeontologica Polonica* 57, no. 2 (2012).

Young, Diana, and Dava Newman. "Augmenting Exploration: Aerospace, Earth and Self." In *Wearable Monitoring Systems*, edited by A. Bonfiglio and D. De Rossi. Boston: Springer, 2011.

6: Bonding With Nature

Brown, Valeria J. "Better Bonding With Beans." *Environmental Health Perspectives* 113, no. 8 (August 2005): A538–41.

Carrington, Emily. Interview with author, December 3, 2020.

———. "Seasonal Variation in the Attachment Strength of Blue Mussels: Causes and Consequences." *Limnology and Oceanography* 47, no. 6 (2002): 1723–33.

Cohen, Noy, J. Herbert Waite, Robert M. McMeeking, and Megan T. Valentine. "Force Distribution and Multiscale Mechanics in the Mussel Byssus." *Philosophical Transactions of the Royal Society B* 374, no. 1784 (2019). https://doi.org/10.1098/rstb.2019.0202.

Fleur, N. S. "Starting Fires to Unearth How Neanderthals Made Glue." *New York Times*, September 7, 2017. https://www.nytimes.com/2017/09/07/science/neanderthals-tar-glue.html.

Frihart, C. R., and L. F. Lorenz. "Protein Adhesives." In *Handbook of Adhesive Technology*, 3rd ed., edited by A. Pizzi and K. L. Mittal, 145–75, Boca Raton, FL: CRC Press, 2018.

Gross, M. "Getting Stuck In." *Chemistry World*, December 2011. https://www.rsc.org/images/Bioadhesives%20-%20Getting%20Stuck%20In_tcm18-210693.pdf.

Hautala, Keith. "We Make Innovations That Stick: Sustainable Adhesives." Oregon State University, December 18, 2019. https://cbee.oregonstate.edu/node/825.

Holten-Andersen, N., Matthew J. Harrington, Henrik Birkedal, Bruce P. Lee, Phillip B. Messersmith, Ka Yee C. Lee, and J. Herbert Waite. "pH-Induced Metal-Ligand Cross-Links Inspired by Mussel Yield Self-Healing Polymer Networks With Near-Covalent Elastic Moduli." *Proceedings of the National Academy of Sciences* 108, no. 7 (2011): 2651–55. https://doi.org/10.1073/pnas.1015862108.

Hopkin, M. "Bacterium Makes Nature's Strongest Glue." *Nature*, April 10, 2006. https://www.nature.com/news/2006/060410/full/news060410-1.html.

Houtman, Nick. "Oregon State University Researcher Receives National Award for Soy-Based Adhesive." Oregon State University, November 8, 2017. https://today.oregonstate.edu/news/oregon-state-university-researcher-receives-national-award-soy-based-adhesive.

IMARC. "Plywood Market: Global Industry Trends, Share, Size, Growth, Opportunity and Forecast 2021–2026." Press release. https://www.imarcgroup.com/plywood-market.

Kotta, J., M. Futter, A. Kaasik, K. Liversage, M. Rätsep, F. R. Barboza, L. Bergström, et al. "Cleaning Up Seas Using Blue Growth Initiatives: Mussel Farming for Eutrophication Control in the Baltic Sea." *Science of the Total Environment* 709 (2020): 136144.

Lambert, Thomas. *Bone Products and Manures: A Treatise on the Manufacture of Fat, Glue, Animal Charcoal, Size, Gelatin, and Manures.* London: Scott Greenwood, 1925.

Lanksbury, J., B. Lubliner, M. Langness, and J. West. "Stormwater Action Monitoring 2015/16 Mussel Monitoring Survey: Final Report." Washington Department of Fish and Wildlife, August 9, 2017. https://wdfw.wa.gov/publications/01925.

Lee, Bruce P., P. B. Messersmith, J. N. Israelachvili, and J. H. Waite. "Mussel-Inspired Adhesives and Coatings." *Annual Review of Materials Research* 41, no. 1 (2011): 99–132. https://doi.org/10.1146/annurev-matsci-062910-100429.

Li, J., C. Green, A. Reynolds, H. Shi, and J. M. Rotchell. "Microplastics in Mussels Sampled From Coastal Waters and Supermarkets in the United Kingdom." *Environmental Pollution* 241 (2018): 35–44.

Li, K. "Biomimicry Case Study—Purebond® Technology: Wood Glue Without Formaldehyde." Biomimicry Institute, n.d. http://toolbox.biomimicry.org/wp-content/uploads/2016/03/CS_PureBond_TBI_Toolbox-2.pdf.

Liu, Y., and K. Li. "Chemical Modification of Soy Protein for Wood

Adhesives." *Macromolecular Rapid Communications* 23, no. 13 (2002): 739–42.

National Cancer Institute. "Formaldehyde and Cancer: Questions and Answers." 2004. https://permanent.fdlp.gov/lps100006/www.cancer.gov/images/Documents/687f2693-82b5-4ec7-9c6f-e4e917d6ee53/fs3_8.pdf.

Needham, Jill, and Kathleen Alcalá. "Mussel Strength: How Mussels Serve Our Ecosystem." *Salish Magazine*, n.d. https://salishmagazine.org/mussel-strength/.

O'Donnell, M. J., M. N. George, and E. Carrington. "Mussel Byssus Attachment Weakened by Ocean Acidification." *Nature Climate Change* 3, no. 6 (2013): 587–90.

Ornes, S. "Mussels' Sticky Feet Lead to Applications." *Proceedings of the National Academy of Sciences* 110, no. 42 (2013): 16697–99.

United Press International. "Mussel's 'Super Glue' May Help Healing." July 7, 1983. https://upi.com/5302609.

United Soybean Board. "Soy-Based Wood Adhesives." 2012. https://soynewuses.org/wp-content/uploads/44422_TDR_Adhesives.pdf.

von Byern, J., and I. Grunwald. *Biological Adhesive Systems. From Nature to Technical and Medical Application*. Springer Science and Business Media, 2010.

Waite, J. Herbert. Interview with author, December 9, 2020.

———. "Mussel Power." *Nature Materials* 7 (2008): 8–9. https://doi.org/10.1038/nmat2087.

———. "Nature's Underwater Adhesive Specialist." *International Journal of Adhesion and Adhesives* 7, no. 1 (1987): 9–14. https://doi.org/10.1016/0143-7496(87)90048-0.

Waite, J. Herbert, and Matthew James Harrington. "Following the Thread: *Mytilus* Mussel Byssus as an Inspired Multi-Functional Biomaterial." *Canadian Journal of Chemistry*, October 28, 2021. https://doi.org/10.1139/cjc-2021-0191.

Waite, J. Herbert, Niels Holten-Andersen, Scott Jewhurst, and Chengjun Sun. "Mussel Adhesion: Finding the Tricks Worth Mimicking." *Journal of Adhesion* 81, no. 3–4 (2005): 297–317. https://doi.org/10.1080/00218460590944602.

Wiggins, Glenn B. *Caddisflies: The Underwater Architects*. Toronto: University of Toronto Press, 2004.

7: Concrete Evidence

Akpan, Nsikan, and Matt Erichs. "Want to Cut Carbon Emissions? Try Growing Cement Bricks With Bacteria." *PBS Newshour*, March 7, 2017. https://www.pbs.org/newshour/science/carbon-emissions-growing-cement-bricks-bacteria-biomason.

Armstrong, S. "These Startups Are Turning CO_2 Pollution Into Something Useful." *Wired*, September 4, 2018. https://www.wired.co.uk/article/xprize-global-warming-climate-change-co2-pollution.

Biomason. "Revolutionizing Cement With Biotechnology." https://www.biomason.com.

Blue Planet Systems. https://www.blueplanetsystems.com/technology.

Breakthrough Energy. https://www.breakthroughenergy.org.

CarbonCure. "Reducing Carbon, One Truck at a Time." https://www.carboncure.com.

Feldman, Amy. "Startup Biomason Makes Biocement Tiles, Retailer H&M Group Plans to Outfit Its Stores' Floors With Them." *Forbes*, June 14, 2021. https://www.forbes.com/sites/amyfeldman/2021/06/14/startup-biomason-makes-bio-cement-tiles-retailer-hm-group-plans-to-outfit-its-stores-floors-with-them/?sh=4908dec257c9.

Fong, P., and V. J. Paul. "Coral Reef Algae." In *Coral Reefs: An Ecosystem in Transition*, edited by Zvy Dubinsky and Noga Stambler, 241–72. New York: Springer, 2011.

Gates, Bill. "Buildings Are Bad for the Climate." *GatesNotes* (blog), October 2019. https://www.gatesnotes.com/Energy/Buildings-are-good-for-people-and-bad-for-the-climate.

———. *How to Avoid a Climate Disaster.* New York: Alfred A. Knopf, 2021.

GlobeNewswire. "Global Concrete Market to Generate $972.04 Billion by 2030: Allied Market Research." August 9, 2021. https://www.globenewswire.com/news-release/2021/08/09/2277251/0/en/Global-Concrete-Market-to-Generate-972-04-Billion-by-2030-Allied-Market-Research.html.

Goel, M., and M. Sudhakar. *Carbon Utilization: Applications for the Energy Industry.* Singapore: Springer, 2017.

Gregory, Jeremy, Hessam AzariJafari, Ehsan Vahidi, Fengdi Guo, Franz-Josef Ulm, and Randolph Kirchain. "The Role of Concrete in Life Cycle Greenhouse Gas Emissions of US Buildings and Pavements." *Proceedings of the National Academy of Sciences* 118, no. 37 (2021). https://doi.org/10.1073/pnas.2021936118.

King, Anthony. "System to Rid Space Station of Astronaut Exhalations Inspires Earth-Based CO_2 Removal." *Horizon*, November 12, 2018. https://ec.europa.eu/research-and-innovation/en/horizon-magazine/system-rid-space-station-astronaut-exhalations-inspires-earth-based-co2-removal.

Margolies, J. "Concrete, A Centuries-Old Material, Gets a New Recipe." *New York Times*, August 11, 2020. https://www.nytimes.com/2020/08/11/business/concrete-cement-manufacturing-green-emissions.html.

Miller, S. A., A. Horvath, and P. J. Monteiro. "Impacts of Booming Concrete Production on Water Resources Worldwide." *Nature Sustainability* 1, no. 1 (2018): 69–76.

NASA. "Carbon Capture Process Makes Sustainable Oil." *NASA Spinoff*, 2019. https://spinoff.nasa.gov/Spinoff2019/ee_4.html.

———. "Closing the Loop: Recycling Water and Air in Space." 2004. https://www.nasa.gov/pdf/146558main_RecyclingEDA(final)%204_10_06.pdf.

Nature. "Concrete Needs to Lose Its Colossal Carbon Footprint." Editorial, September 28, 2021. https://www.nature.com/articles/d41586-021-02612-5.

Rosic, Nedeljka, Edmund Yew Siang Ling, Chon-Kit Kenneth Chan, Hong Ching Lee, Paulina Kaniewska, David Edwards, Sophie Dove, and Ove Hoegh-Guldberg. "Unfolding the Secrets of Coral–Algal Symbiosis." *ISME J* 9, (2015): 844–56. https://doi.org/10.1038/ismej.2014.182

Ryan, K. J. "How This Startup Is Using Bacteria to Grow Bricks From Scratch." *Inc.*, 2016. https://www.inc.com/kevin-j-ryan/best-industries-2016-sustainable-building-materials.html.

Shepherd, Bart. Interview with author, November 8, 2018.

Solidia. https://www.solidiatech.com.

Stokstad, E. "Human 'Stuff' Now Outweighs All Life on Earth." *Science*, December 9, 2020. https://www.sciencemag.org/news/2020/12/human-stuff-now-outweighs-all-life-earth.

Sully, S., D. E. Burkepile, M. K. Donovan, G. Hodgson, and R. van Woesik. "A Global Analysis of Coral Bleaching Over the Past Two Decades." *Nat Commun* 10, no. 1264 (2019). https://doi.org/10.1038/s41467-019-09238-2.

Timperley, Jocelyn. "Q&A: Why Cement Emissions Matter for Climate Change." *CarbonBrief*, 2018. https://www.carbonbrief.org/qa-why-cement-emissions-matter-for-climate-change.

XPRIZE Foundation. "From Carbon to Concrete." March 9, 2017. https://www.xprize.org/prizes/carbon/articles/from-carbon-to-concrete.

8: Driving on a Seed

Barboza, Tony. "Court Allows Exide to Abandon a Toxic Site in Vernon. Taxpayers Will Fund the Cleanup." *Los Angeles Times*, October 16, 2020. https://www.latimes.com/california/story/2020-10-16/exide-bankrtuptcy-decision-vernon-cleanup.

Belcher, Angela. "Using Nature to Grow Batteries." TEDx Talk, April 2011. https://www.ted.com/talks/angela_belcher_using_nature_to_grow_batteries/transcript?language=en.

BloombergNEF. "Electric Vehicle Sales to Fall 18% in 2020 but Long-term Prospects Remain Undimmed." May 19, 2020. https://about.bnef.com/blog/electric-vehicle-sales-to-fall-18-in-2020-but-long-term-prospects-remain-undimmed/.

Chen, P., Y. Wu, Y. Zhang, T.-H. Wu, Y. Ma, C. Pelkowski, H. Yang, Y. Zhang, X. Hu, and N. Liu. "A Deeply Rechargeable Zinc Anode With Pomegranate-Inspired Nanostructure for High-Energy Aqueous Batteries." *Journal of Materials Chemistry A* 6, no. 44 (2018): 21933–40.

Chu, S., Y. Cui, and N. Liu. "The Path Towards Sustainable Energy." *Nature Materials* 16, no. 1 (2017): 16–22.

Crabtree, G. "Perspective: The Energy-Storage Revolution." *Nature* 526, no. 7575 (2015): S92.

Cui, Y. "New 'Pomegranate-Inspired' Design Solves Problems for Lithium-Ion Batteries." SLAC National Accelerator Laboratory, February

2014. https://www6.slac.stanford.edu/news/2014-02-16-pomegranate-inspired-batteries.aspx.

Dalton, P. J., E. Bowens, T. North, S. Balcer, and A. Rocketdyne. "International Space Station Lithium-Ion Battery Status." Presented at the NASA Aerospace Battery Workshop, November 2019.

Fagan, Mary. "Sheikh Yamani Predicts Price Crash as Age Oil Ends." *Telegraph*, June 25, 2000. https://www.telegraph.co.uk/news/uknews/1344832/Sheikh-Yamani-predicts-price-crash-as-age-of-oil-ends.html.

Fletcher, S. *Bottled Lightning: Superbatteries, Electric Cars, and the New Lithium Economy.* New York: Hill and Wang, 2011.

Galbraith, K. "Charging Ahead." *Stanford Magazine*, July/August 2014. https://stanfordmag.org/contents/charging-ahead.

Garcia, M. "Spacewalkers Complete Multi-Year Effort to Upgrade Space Station Batteries." NASA, February 1, 2021. https://www.nasa.gov/feature/spacewalkers-complete-multi-year-effort-to-upgrade-space-station-batteries.

Gordon, Robert J. *The Rise and Fall of American Growth.* New Jersey: Princeton University Press, 2017.

Han, X., L. Lu, Y. Zheng, X. Feng, Z. Li, J. Li, and M. Ouyang. "A Review on the Key Issues of the Lithium Ion Battery Degradation Among the Whole Life Cycle." *eTransportation* 1 (2019): 100005.

Iida, Fumiya. Interview with author, March 16, 2021.

Khan Academy. "Are Viruses Dead or Alive?" N.d. https://www.khanacademy.org/test-prep/mcat/cells/viruses/a/are-viruses-dead-or-alive.

Kotov, Nicholas. Interview with author, March 11, 2021.

Lee, Y. J., H. Yi, W. J. Kim, K. Kang, D. S. Yun, M. S. Strano, G. Ceder, and A. M. Belcher. "Fabricating Genetically Engineered High-Power Lithium-Ion Batteries Using Multiple Virus Genes." *Science* 324, no. 5930 (April 2, 2009): 1051–55. https://www.science.org/doi/10.1126/science.1171541.

Li, W., Z. Liang, Z. Lu, H. Yao, Z. W. Seh, K. Yan, G. Zheng, and Y. Cui. "A Sulfur Cathode With Pomegranate-Like Cluster Structure." *Advanced Energy Materials* 5, no. 16 (2015). https://web.stanford.edu/group/cui_group/papers/Weiyang_Cui_AEM_2015.pdf.

Liu, N., Z. Lu, J. Zhao, M. T. McDowell, H.-W. Lee, W. Zhao, and Y. Cui. "A Pomegranate-Inspired Nanoscale Design for Large-Volume-Change Lithium Battery Anodes." *Nature Nanotechnology* 9, no. 3 (2014): 187–92.

Nam, K. T., D.-W. Kim, P. J. Yoo, C.-Y. Chiang, N. Meethong, P. T. Hammond, Y.-M. Chiang, and A. M. Belcher. "Virus-Enabled Synthesis and Assembly of Nanowires for Lithium Ion Battery Electrodes." *Science* 312, no. 5775 (2006): 885–88.

Oberhaus, Daniel. "The Next Generation of Batteries Could Be Built by Viruses." *Wired*, February 26, 2020. https://www.wired.com/story/the-next-generation-of-batteries-could-be-built-by-viruses/.

Oh, D., J. Qi, B. Han, G. Zhang, T. J. Carney, J. Ohmura, Y. Zhang, Y. Shao-Horn, and A. M. Belcher. "M13 Virus-Directed Synthesis of Nanostructured Metal Oxides for Lithium–Oxygen Batteries." *Nano Letters* 14, no. 8 (2014): 4837–45.

Pearce, Fred. "Getting the Lead Out: Why Battery Recycling Is a Global Health Hazard." *Yale Environment 360*, November 2, 2020. https://e360.yale.edu/features/getting-the-lead-out-why-battery-recycling-is-a-global-health-hazard.

Royal Swedish Academy of Sciences. "Lithium-Ion Batteries." Scientific background on the Nobel Prize in Chemistry 2019. https://www.nobelprize.org/uploads/2019/10/advanced-chemistryprize2019.pdf.

Rutgers School of Arts and Sciences. "Edison and the Electric Car." Thomas A. Edison Papers, October 28, 2016. http://edison.rutgers.edu/elecar.htm.

Service, R. F. "How to Build a Better Battery Through Nanotechnology." *Science*, May 26, 2016. https://www.sciencemag.org/news/2016/05/how-build-better-battery-through-nanotechnology.

University of Michigan. "Nicholas A. Kotov." Chemical Engineering faculty bio. https://che.engin.umich.edu/people/kotov-nicholas/.

Wang, M., D. Vecchio, C. Wang, A. Emre, X. Xiao, Z. Jiang, P. Bogdan, Y. Huang, and N. A. Kotov. "Biomorphic Structural Batteries for Robotics." *Science Robotics* 5, no. 45 (2020).

Xu, K. "A Long Journey of Lithium: From the Big Bang to Our Smartphones." *Energy & Environmental Materials* 2, no. 4 (2019): 229–33.

9: Skeletons in the Closet

Altair. Customer Stories. https://www.altair.com/resourcelibrary/?category=Customer+Stories.

———. "The Implant Boom: It's Now Hip to Replace Your Hip." June 18, 2018. https://www.altair. com/newsroom/articles/implant-boom-hip-to-replace-your-hip/?mc_cid=f09a559bcd&mc_eid= 8d0667d553.

———. "Lushan Primary School: Using Remote Robotic Construction Techniques for Architectural Development of an Extraordinary School in Rural China." N.d. https://www.altair.com/customer-story/zaha-hadid-lushan-primary-school.

Bhatia, Aatish. "What Your Bones Have in Common With the Eiffel Tower." *Wired*, March 9, 2015. https://www.wired.com/2015/03/empzeal-eiffel-tower/.

Brennan, Jeff. Interview with author, October 12, 2020.

Buenzli, P. R., and N. A. Sims. "Quantifying the Osteocyte Network in the Human Skeleton." *Bone* 75 (2015): 144–50.

da Vinci, Leonardo. *Leonardo da Vinci's Notebooks: Arranged and Rendered Into English With Introductions*. Trans. Edward McCurdy. London: Duckworth and Company, 1908.

Donahue, S. W., S. J. Wojda, M. E. McGee-Lawrence, J. Auger, and H. L. Black. "Osteoporosis Prevention in an Extraordinary Hibernating Bear." *Bone* 145 (2021): 115845.

Doube, M., M. M. Klosowski, A. M. Wiktorowicz-Conroy, J. R. Hutchinson, and S. J. Shefelbine.

"Trabecular Bone Scales Allometrically in Mammals and Birds." *Proceedings of the Royal Society B: Biological Sciences* 278, no. 1721 (2011): 3067–73.

Eiffel Tower tourism website. "Origins and Construction of the Eiffel Tower." https://www.toureiffel.paris/en/the-monument/history.

Galilei, Galileo. *Dialogues Concerning Two New Sciences*. Trans. Alfonso de Salvio and Henry Crew. New York: Macmillan Company, 1914.

Haldane, J. B. S. "On Being the Right Size." 1926. https://www.phys.ufl.edu/courses/phy3221/spring10/HaldaneRightSize.pdf.

Hermanussen, M., C. Scheffler, D. Groth, and C. Aßmann. "Height and Skeletal Morphology in Relation to Modern Life Style." *Journal of Physiological Anthropology* 34, no. 1 (2015): 1–5.

Johnson, G. "Of Mice and Elephants: A Matter of Scale." *New York Times*, January 12, 1999. http://hep.ucsb.edu/courses/ph6b_99/0111299sci-scaling.html.

McGee-Lawrence, M., P. Buckendahl, C. Carpenter, K. Henriksen, M. Vaughan, and S. Donahue. "Suppressed Bone Remodeling in Black Bears Conserves Energy and Bone Mass During Hibernation." *Journal of Experimental Biology* 218, no. 13 (2015): 2067–74.

NASA. "Preventing Bone Loss in Spaceflight With Prophylactic Use of Bisphosphonate: Health Promotion of the Elderly by Space Medicine Technologies." March 27, 2019. https://www.nasa.gov/mission_pages/station/research/news/b4h-3rd/hh-preventing-bone-loss-in-space.

National Institutes of Health. "What Is Bone?" October 2018. https://www.bones.nih.gov/health-info/bone/bone-health/what-is-bone.

Ryan, T. M., and C. N. Shaw. "Gracility of the Modern Homo Sapiens Skeleton Is the Result of Decreased Biomechanical Loading." *Proceedings of the National Academy of Sciences* 112, no. 2 (2015): 372–77.

Sozen, T., L. Özışık, and N. Ç. Başaran. "An Overview and Management of Osteoporosis." *European Journal of Rheumatology* 4, no. 1 (2017): 46.

Vogel, S. "Cats' Paws and Catapults: Mechanical Worlds of Nature and People." New York: W. W. Norton, 2000.

Wade, N. "Your Body Is Younger Than You Think." *New York Times*, August 2, 2005, https://www.nytimes.com/2005/08/02/science/your-body-is-younger-than-you-think.html.

Wagner, D. O., and P. Aspenberg. "Where Did Bone Come From? An Overview of Its Evolution." *Acta Orthopaedica* 82, no. 4 (2011): 393–98.

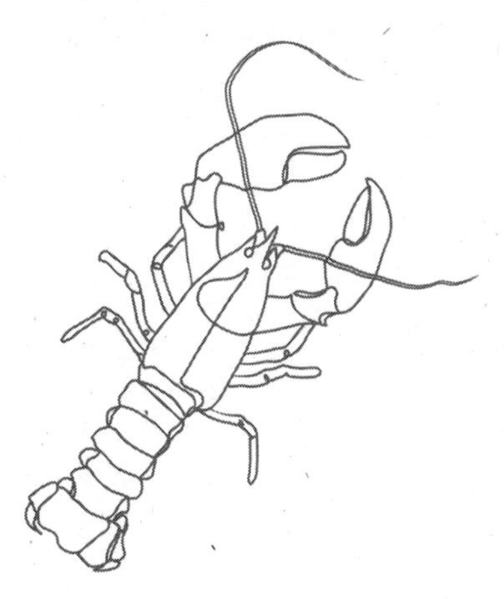

10: A Monster Is Born

Amylin. "The Discovery and Development of Byetta (Exenatide) Injection and Bydureon (Exenatide Extended-Release for Injectable Suspension)." 2012. https://www.multivu.com/assets/53897/documents/53897-Exenatide-History-FINAL-original.pdf.

Beck, D. D., B. E. Martin, and C. H. Lowe. *Biology of Gila Monsters and Beaded Lizards*. Vol. 9. Berkeley: University of California Press, 2005.

Bordon, Karla de Castro Figueiredo, Camila Takeno Cologna, Elisa Corrêa Fornari-Baldo, Ernesto Lopes Pinheiro-Júnior, Felipe Augusto Cerni, Fernanda Gobbi Amorim, Fernando Antonio Pino Anjolette, et al. "From Animal Poisons and Venoms to Medicines: Achievements, Challenges and Perspectives in Drug Discovery." *Frontiers in Pharmacology* 11 (July 24, 2020): 1132. https://pubmed.ncbi.nlm.nih.gov/32848750/.

Bridges, A., K. G. Bistas, and T. F. Jacobs. "Exenatide." In *StatPearls [Internet]*. Treasure Island, FL: StatPearls Publishing, July 19, 2021. https://www.ncbi.nlm.nih.gov/books/NBK518981/

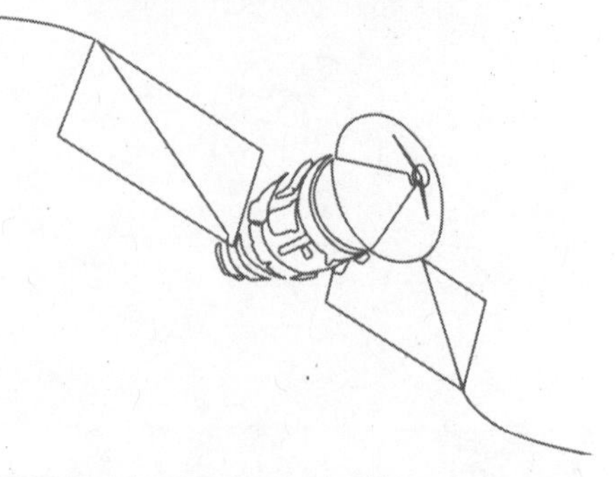

Calcabrini, C., E. Catanzaro, A. Bishayee, E. Turrini, and C. Fimognari. "Marine Sponge Natural Products With Anticancer Potential: An Updated Review." *Marine Drugs* 15, no. 10 (2017): 310.

DeNardo, Dale. Interview with author, June 2, 2020.

Emerson, Ralph Waldo. *The Prose Works of Ralph Waldo Emerson: Representative Men. English Traits. Conduct of Life.* Boston: J. R. Osgood and Company, 1872.

Gellene, Denise. "Lizard Is Source of Newest Diabetes Drug." *Los Angeles Times*, April 30, 2005. https://www.latimes.com/archives/la-xpm-2005-apr-30-fi-gila30-story.html.

Goodfellow, George. "The Gila Monster Again." *Scientific American* 96, no. 13 (1907): 271. http://www.jstor.org/stable/26005510.

Holst, J. J., and J. Gromada. "Role of Incretin Hormones in the Regulation of Insulin Secretion in Diabetic and Nondiabetic Humans." *Am J Physiol Endocrinol Metab* 287, no. 2 (2004): E199–205.

Kalinyak, Judith. Interview with author, June 20, 2021.

Lazarovici, Philip, Cezary Marcinkiewicz, and Peter I. Lelkes. "From Snake Venom's Disintegrins and C-Type Lectins to Anti-Platelet Drugs." *Toxins* 11, no. 5 (May 27, 2019): 303. https://www.mdpi.com/2072-6651/11/5/303.

Nation, Earl F. "George E. Goodfellow, M.D. (1855–1910): Gunfighter's Surgeon and Urologist." *Urology* 2, no. 1 (1973): 85–92. https://doi.org/10.1016/0090-4295(73)90226-4.

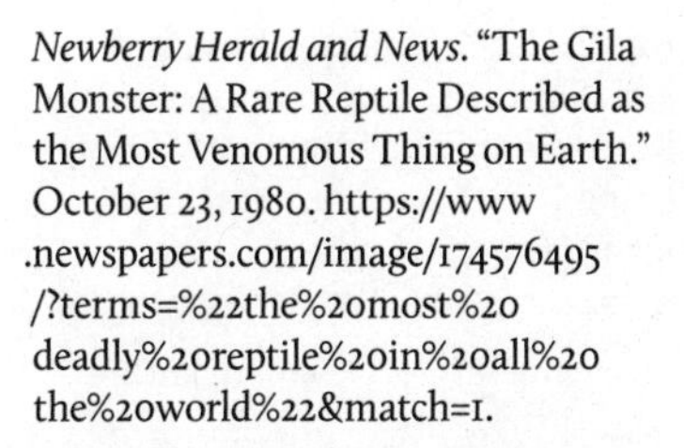

Newberry Herald and News. "The Gila Monster: A Rare Reptile Described as the Most Venomous Thing on Earth." October 23, 1980. https://www.newspapers.com/image/174576495/?terms=%22the%20most%20deadly%20reptile%20in%20all%20the%20world%22&match=1.

NIH National Institute on Aging. "Exendin-4: From Lizard to Laboratory… and Beyond." July 11, 2012. https://www.nia.nih.gov/news/exendin-4-lizard-laboratory-and-beyond.

Nunez, Christina. "Deforestation Explained." *National Geographic*, November 2, 2021. https://www.nationalgeographic.com/environment/article/deforestation.

Piper, Ross, Alexander Kagansky, John Malone, Nils Bunnefeld, and Rob Jenkins. "Nature Is a Rich Source of Medicine—If We Can Protect It." *The Conversation*, December 13, 2018. https://theconversation.com/nature-is-a-rich-source-of-medicine-if-we-can-protect-it-107471.

Pollack, Andrew. "Lizard-Linked Therapy Has Roots in the Bronx." *New York Times*, September 21, 2002. https://www.nytimes.com/2002/09/21/business/lizard-linked-therapy-has-roots-in-the-bronx.html.

San Francisco Chronicle. "Ugly as Sin Itself." June 18, 1893. https://www.newspapers.com/image/27601356/?terms=Ugly%20as%20Sin%20Itself%20gila%20monster.

Sengupta, Amitdyuti, and Jagadananda Behera. "Compre-hensive View on Chemistry, Manufacturing and Applications of Lanolin Extracted From Wool Pretreatment." *American Journal of Engineering Research* 3, no. 7 (2014): 33–43.

Tan, Siang Yong, and Jason Merchant. "Frederick Banting (1891-1941): Discoverer of Insulin." *Singapore Medical Journal* 58, no. 1 (2017): 2–3. https://doi.org/10.11622/smedj.2017002.

Tan, Siang Yong, and Y. Tatsumura. "Alexander Fleming (1881–1955): Discoverer of Penicillin." *Singapore Medical Journal* 56, no. 7 (2015): 366–67. https://doi.org/10.11622/smedj.2015105.

Tyron, George Washington, Henry Augustus Pilsbry, and B. Sharp. *Manual of Conchology, Structural and Systematic.* Vol. 6, *Conidae, Pleurotomidae.* Philadelphia: Academy of Natural Sciences, 1884.

Wootson, Cleve, Jr. "'I Guess You Are Here for the Opium': Investigator Stumbles Across $500 Million in Poppy Plants." *Washington Post*, May 25, 2017. https://www.washingtonpost.com/news/to-your-health/wp/2017/05/25/i-guess-you-are-here-for-the-opium-investigator-stumbles-across-500-million-in-poppy-plants/.

World Health Organization. "The Worldwide Rise of Chronic Non-communicable Diseases: A Slow-Motion Catastrophe." Opening remarks at the First Global Ministerial Conference on Healthy Lifestyles and Noncommunicable Disease Control. April 28, 2011. https://www.who.int/director-general/speeches/detail/the-worldwide-rise-of-chronic-noncommunicable-diseases-a-slow-motion-catastrophe.

11: Bumps Are Beautiful

Bernitsas, M. M., and K. Raghavan. Converter of Current, Tide, or Wave Energy. European Patent EP 1 812 709 B1, issued on April 17, 2013.

———. Fluid Motion Energy Converter. U.S. Patent 7,493,759 B2, issued on February 24, 2009.

Bernitsas, M. M., K. Raghavan, Y. Ben-Simon, and E. Garcia. "VIVACE (Vortex Induced Vibration Aquatic Clean Energy): A New Concept in Generation of Clean and Renewable Energy From Fluid Flow." *J Offshore Mech Arct Eng* 130, no. 4 (2008): 041101.

Bernitsas, Michael, and Tad Dritz. "Low Head, Vortex Induced Vibrations River Energy Converter." Office of Scientific and Technical Information (OSTI), U.S. Department of Energy, June 30, 2006. https://www.osti.gov/servlets/purl/896401.

Darwin, Charles. *On the Origin of Species*. London: John Murray, 1859.

Fish, F., and G. V. Lauder. "Passive and Active Flow Control by Swimming Fishes and Mammals." *Annu Rev Fluid Mech* 38 (2006): 193–224.

Fish, F. E., P. W. Weber, M. M. Murray, and L. E. Howle. "The Tubercles on Humpback Whales' Flippers: Application of Bio-Inspired Technology." *Integrative and Comparative Biology* 51, no. 1 (July 2011): 203–13.

Fish, Frank. Interview with author, August 26, 2020.

Gierach, John. *Trout Bum*. Berkeley: West Margin Press, 2013.

Kim, G. Y., C. Lim, E. S. Kim, and S. C. Shin. "Prediction of Dynamic Responses of Flow-Induced Vibration Using Deep Learning." *Appl Sci* 11, no. 15 (2021): 7163. https://www.mdpi.com/2076-3417/11/15/7163.

Liao, James. Interview with author, August 19, 2020.

Liu, Yuqiang, Na Sun, Jiawei Liu, Zhen Wen, Xuhui Sun, Shuit-Tong Lee, and Baoquan Sun. "Integrating a Silicon Solar Cell With a Triboelectric Nanogenerator via a Mutual Electrode for Harvesting Energy From Sunlight and Raindrops." ACS *Nano* 12, no. 3 (2018): 2893–99. https://doi.org/10.1021/acsnano.8b00416.

National Institute of Standards and Technology (NIST). https://www.nist.gov/.

Raghavan, K., and M. M. Bernitsas. "Experimental Investigation of Reynolds Number Effect on Vortex Induced Vibration of Rigid Cylinder on Elastic Supports." *Ocean Engineering* 38, no. 5–6 (April 2011): 719–31.

United Nations. "Achieving Targets on Energy Helps Meet Other Global Goals, UN Forum Told." Sustainable Development Goals, July 11, 2018. https://www.un.org/sustainabledevelopment/blog/2018/07/achieving-targets-on-energy-helps-meet-other-global-goals-un-forum-told-2/.

University of Michigan. "Michael Bernitsas." Naval Architecture and Marine Engineering faculty bio, n.d. https://name.engin.umich.edu/people/bernitsas-michael/.

Wilford, John Noble. "How the Whale Lost Its Legs and Returned to the Sea." *New York Times*, May 3, 1994. https://www.nytimes.com/1994/05/03/science/how-the-whale-lost-its-legs-and-returned-to-the-sea.html.

12: Window Pain

"Birds Eat 400 to 500 Million Tons of Insects Annually." *ScienceDaily*, July 8, 2018. https://www.sciencedaily.com/releases/2018/07/180709100850.htm.

Blackledge, T. A., and J. W. Wenzel. "Do Stabilimenta in Orb Webs Attract Prey or Defend Spiders?" *Behavioral Ecology* 10, no. 4 (1999): 372–76.

Brown, B. B., L. Hunter, and S. Santos. "Bird-Window Collisions: Different Fall and Winter Risk and Protective Factors." *PeerJ* 8 (2020): e9401.

Bruce, M. J., A. M. Heiling, and M. E. Herberstein. "Spider Signals: Are Web Decorations Visible to Birds and Bees?" *Biology Letters* 1, no. 3 (2005): 299–302.

Chu, Jennifer. "Super-Strong Surgical Tape Detaches on Demand." *MIT News*, June 22, 2020. https://news.mit.edu/2020/surgical-tape-wounds-0622.

DeGroote, Luke. Interview with author, September 1, 2020.

Gruener, Anna. "The Effect of Cataracts and Cataract Surgery on Claude Monet." *The British Journal of General Practice: The Journal of the Royal College of General Practitioners* 65, no. 634 (2015): 254–55. https://bjgp.org/content/65/634/254.

Guo, Qi, Zhujun Shi, Yao-Wei Huang, Emma Alexander, Cheng-Wei Qiu, Federico Capasso, and Todd Zickler. "Compact Single-Shot Metalens Depth Sensors Inspired by Eyes of Jumping Spiders." *Proceedings of the National Academy of Sciences* 116, no. 46 (2019): 22959–65. https://doi.org/10.1073/pnas.1912154116.

Hastad, O., and A. Odeen. "A Vision Physiological Estimation of Ultraviolet Window Marking Visibility to Birds." *PeerJ* 2 (2014): e621.

Hicks, L. "These Frightening Ogre-Faced Spiders Use Their Legs to 'Hear.'" *Science*, October 29, 2020. https://www.sciencemag.org/news/2020/10/these-frightening-ogre-faced-spiders-use-their-legs-hear.

Loss, Scott R., Tom Will, Sara S. Loss, and Peter P. Marra. "Bird–Building Collisions in the United States: Estimates of Annual Mortality and Species Vulnerability." *The Condor* 116, no. 1 (February 1, 2014): 8–23. https://doi.org/10.1650/CONDOR-13-090.1

Martin, Graham. Interview with author, August 27, 2020.

———. *The Sensory Ecology of Birds*. Oxford: Oxford University Press, 2017.

McCray, W. Patrick. *Giant Telescopes: Astronomical Ambition and the Promise of Technology*. Boston: Harvard University Press, 2004.

Nyffeler, M., Ç. H. Sekercioglu, and C. J. Whelan. "Insectivorous Birds Consume an Estimated 400–500 Million Tons of Prey Annually." *The Science of Nature* 105, no. 7 (2018): 1–13.

Ornilux Bird Protection Glass (brochure). http://www.ornilux.com

/Attachments/Project_Brochure _FINAL_081613.pdf.

Runwal, Priyanka. "Building Collisions Are a Greater Danger for Some Birds Than Others." *Audubon*, July 9, 2020. https://www.audubon.org/news/building-collisions-are-greater-danger-some-birds-others.

Stafstrom, J. A., G. Menda, E. I. Nitzany, E. A. Hebets, and R. R. Hoy. "Ogre-Faced, Net-Casting Spiders Use Auditory Cues to Detect Airborne Prey." *Current Biology* 30, no. 24 (2020): 5033–39.

Trafton, A. "Double-Sided Tape for Tissues Could Replace Surgical Sutures." *MIT News*, October 30, 2019. https://news.mit.edu/2019/double-sided-tape-tissues-could-replace-surgical-sutures-1030.

Waldorf, Heath. "What the New Bird Friendly Local Law 15 Means to You." *NYREJ*, June 23, 2020. https://nyrej.com/what-the-new-bird-friendly-local-law-15-means-to-you-by-heath-waldorf.

Welch, Lisa. Interview with author, August 28, 2020.

Winton, R. S., N. Ocampo-Peñuela, and N. Cagle. "Geo-Referencing Bird-Window Collisions for Targeted Mitigation." *PeerJ* 6 (2018): e4215.

Yuk, H., C. E. Varela, C. S. Nabzdyk, X. Mao, R. F. Padera, E. T. Roche, and X. Zhao. "Dry Double-Sided Tape for Adhesion of Wet Tissues and Devices." *Nature* 575, no. 7781 (November 2019): 169–74. https://www.nature.com/articles/s41586-019-1710-5.

Zschokke, S. "Ultraviolet Reflectance of Spiders and Their Webs." *Journal of Arachnology* 30, no. 2 (2002): 246–54.

13: Flashes of Brilliance

Chalfie, Martin. "GFP: Lighting Up Life." Nobel lecture, December 8, 2008. https://www.pnas.org/content/106/25/10073.

Chalfie, Martin, Y. Tu, G. Euskirchen, W. W. Ward, and D. C. Prasher. "Green Fluorescent Protein as a Marker for Gene Expression." *Science* 263, no. 5148 (1994): 802–5.

Franklin, Benjamin. *The Works of Dr. Benjamin Franklin: Philosophical. Essays and Correspondence*. London: John Sharpe, 1809.

Grynkiewicz, G., M. Poenie, and R. Y. Tsien. "A New Generation of Ca2+ Indicators With Greatly Improved Fluorescence Properties," *J Biol Chem* 260 (1985): 3440–50.

Harvey, Edmund Newton. *A History of Luminescence: From the Earliest Times Until 1900*. Philadelphia: American Philosophical Society, 1957.

Hodgkin, A. L., and A. F. Huxley. "Action Potentials Recorded From Inside a Nerve Fibre." *Nature* 144 (1939): 710–11.

Hogue, Charles Leonard. *Latin American Insects and Entomology.* Berkeley: University of California Press, 1993.

Karagianis, Liz. "The Brilliance of Basic Research." *MIT Spectrum*, Spring 2014. https://spectrum.mit.edu/spring-2014/the-brilliance-of-basic-research/.

Minta, A., J. P. Y. Kao, and R. Y. Tsien. "Fluorescent Indicators for Cytosolic Calcium Based on Rhodamine and Fluorescein Chromophores." *J Biol Chem* 264 (1989): 8171–78.

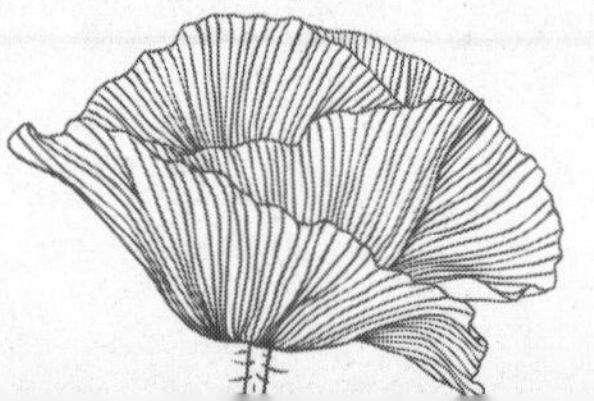

Nobel Foundation. "Glowing Proteins—A Guiding Star for Biochemistry." Nobel Prize in Chemistry 2008 press release, October 8, 2008. https://www.nobelprize.org/prizes/chemistry/2008/press-release/.

———. "Martin Chalfie: Facts." https://www.nobelprize.org/prizes/chemistry/2008/chalfie/facts/.

———. "Roger Y. Tsien: Facts." https://www.nobelprize.org/prizes/chemistry/2008/tsien/facts/.

Pieribone, V., and D. F. Gruber. *Aglow in the Dark: The Revolutionary Science of Biofluorescence*. Boston: Harvard University Press, 2005.

Pliny the Elder. *The Natural History of Pliny*. Trans. Henry Thomas Riley and John Bostock. Vol. 6. London: George Bell and Sons, 1989.

Prasher, D. C., V. K. Eckenrode, W. W. Ward, F. G. Prendergast, and M. J. Cormier. "Primary Structure of the Aequorea Victoria Green-Fluorescent Protein." *Gene* 111, no. 2 (February 15, 1992): 229–33.

Schwiening, Christof J. "A Brief Historical Perspective: Hodgkin and Huxley." *Journal of Physiology* 590, no. 11 (2012): 2571–75.

Shimomura, O. "The Discovery of Aequorin and Green Fluorescent Protein." *Journal of Microscopy* 217, no. 1 (200): 3–15.

———. "Discovery of Green Fluorescent Protein (GFP) (Nobel Lecture)." *Angewandte Chemie International Edition* 48, no. 31 (2009): 5590–5602.

———. "A Short Story of Aequorin." *Biological Bulletin* 189, no. 1 (1995): 1–5.

Shimomura, O., F. H. Johnson, and Y. Saiga. "Extraction, Purification and Properties of Aequorin, a Bioluminescent Protein from the Luminous Hydromedusan, Aequorea." *Journal of Cellular and Comparative Physiology* 59, no. 3 (1962), 223–39.

Shimomura, O., S. Shimomura, and J. H. Brinegar. *Luminous Pursuit: Jellyfish, GFP, and the Unforeseen Path to the Nobel Prize*. Singapore: World Scientific, 2017.

Tsien, Roger. "Unlocking Cell Secrets With Light Beams and Molecular Spies." Acceptance speech, Heineken Prize for Biochemistry and Biophysics, 2002.

University of California San Diego Health. "2008 Nobel Prize in Chemistry Shared by UC San Diego Researcher Roger Tsien." Press release, October 8, 2008.

Index